The
Developmental
Neurotoxicity
of
Lead

The Developmental Neurotoxicity of Lead

Christopher Winder

Department of Histopathology
Royal Postgraduate Medical School
London

MTP PRESS LIMITED
a member of the KLUWER ACADEMIC PUBLISHERS GROUP
LANCASTER / BOSTON / THE HAGUE / DORDRECHT

Published in the UK and Europe by
MTP Press Limited
Falcon House
Lancaster, England

British Library Cataloguing in Publication Data

Winder, Christopher
  The developmental neurotoxicity of lead.
  1. Lead-poisoning   2. Lead—Environmental
  aspects
  I. Title
  613'.1   RA1231.L4

  ISBN-13:978-94-010-8966-1   e-ISBN-13:978-94-009-5594-3
  DOI: 10.1007/978-94-009-5594-3

Published in the USA by
MTP Press
A division of Kluwer Boston Inc
190 Old Derby Street
Hingham, MA 02043, USA

Library of Congress Cataloguing in Publication Data

Winder, Christopher.
  The developmental neurotoxicity of lead.

  Bibliography: p.
  Includes index.
  1. Nervous system—Diseases.   2. Lead—Toxicology.
  3. Neurotoxic agents.   4. Developmental neurology.
  I. Title.
  RC 346.W574   1984   616.8   84–3877

Phototypeset by Blackpool Typesetting Services Ltd., Blackpool, England

Contents

Acknowledgments   vi

1   Introduction   1

2   Lead and man   17

3   Experimental models of lead administration   33

4   The behavioural effects of lead   53

5   The neurochemical effects of lead   75

6   The morphological effects of lead   101

7   Discussion   123

8   References   131

    Index   155

Acknowledgements

I am indebted to my friends, colleagues and associates for considerable help given to me during the preparation of this book. Particularly, I would like to thank Paul Lewis, Ian Kitchen, Neil Carmichael and Lori Garten for invaluable criticism and discussion of this work. Such merits that this book has are due in part to their efforts, while any defects that remain are the sole responsibility of the author.

Considerable assistance is also gratefully acknowledged from Ruth Powell, Paul Willner, Tim Jordan, Graham Hollins, Steve Lewis and Sebastion Lazareno. Material in this book is reprinted by permission of the University of London and Figure 1 is reprinted by permission of the Trustees of The British Museum. Financial support was received from grants from the Wellcome Foundation and the Health and Safety Executive.

Finally, I would like to thank my wife Judy for her continuing love and support.

Christopher Winder

1
Introduction

Over the past two millennia environmental lead levels have risen dramatically (Patterson, 1980). Most of this increase has occurred since the beginning of the present century (Murozami *et al.*, 1969), and taken into perspective, a typical individual living in the industrialized world sustains a lead burden 500 times that of his prehistoric ancestors (Patterson, 1983).

Lead is unique in being an environmental pollutant where the levels deemed clinically toxic are less than an order of magnitude from those that are normally encountered in the population. The clinical limit for lead exposure, 70 μg Pb/100 ml blood, is about three times the 'normal' value (in the range 15–30 μg Pb/100 ml blood). This small difference is remarkable from the toxicological standpoint. The key concept here is control of dispersal. The industrial actions of man bring contact with a wide variety of substances, some of which are poisonous or dangerous. In such cases protective measures are taken to keep hazardous exposure to a minimum. In the production and distribution of lead compounds, stringent regulations are enforced. The exception to this is in the dispersal of lead from vehicle exhausts following purchase from the petrol pump. Perhaps man's long-standing acquaintance with lead in home and industry is responsible for a complacency which until only recently has been unquestioned.

Beginning in the late 1960s, evidence began to accumulate suggesting that lead at relatively low levels of exposure might be causing negative effects on neurobiological function. Suggestions were made that neuropsychiatric impairment occurred at lower exposures than had previously been believed and perhaps even in the 'normal' range. The reality of such changes has not yet been proved by findings contained in published reports in the scientific literature.

Because of the possible social and economic implications of low-level lead effects, government agencies have tried to examine this issue critically, and numerous reviews attempting clarification have been published (see e.g. US NAS, 1972; UK DoE, 1974; EEC, 1975; US EPA, 1977; US CDC, 1978; US NAS, 1978; WHO, 1977; UK DHSS, 1980; Royal Commission, 1983). Also, non-governmental reviews have been published (e.g. Waldron and Stofen,

1

1974; Singhal and Thomas, 1980; Needleman, 1980; Chisholm and O'Hara, 1982). However, the outcome of such studies seems mainly to have been a heightening of the conflicts between economic interests in industry and commerce on one side, and public health bodies and environmental lobby groups on the other. The lead health problem is increasingly seen as occasioning debate between advocates of either continued use or abandonment of alkyllead antiknock additives in domestic petrol. Each side states what appears to be a balanced viewpoint of the situation, but often this viewpoint is a restatement of existing dogmata (see e.g. Conservation Society, 1979; 1981; Associated Octel, 1982; Friends of the Earth, 1982; the Fellowship of Engineering, 1981; CLEAR, 1983).

THE CHEMISTRY OF LEAD

Lead belongs to the group of heavy metals and has physical properties as shown in Table 1.

Table 1 The physical properties of lead

Atomic number	82
Atomic weight	207.19
Principal oxidation numbers	$+2$, $+4$
Density	$11\,340\,\mathrm{kg\,m^{-3}}$
Melting point	$327.4\,°C$
Boiling point	$1744\,°C$

Among group IV elements it is the most electropositive, and has the most stable divalent state. The coordination number for divalent lead compounds ranges from 2 to 7, and for tetravalent ones from 4 to 8. The stereochemistry is usually octahedral or tetrahedral.

The elemental symbol for lead, Pb, is from the latin *plumbum*, and the alchemical symbol for lead was Saturn. The medieval alchemist considered lead to be the father of metals due to the ease in which the nobler metals, such as silver and gold, dissolved into molten lead. Plumbism and Saturnism are both terms which have been used as synonyms for clinical lead poisoning.

The principle ore of lead is galena, (lead II sulphide, PbS), found in many parts of the world, especially Australia, USA, Spain and Mexico. Other lead minerals, such as anglesite (lead II sulphate, $PbSO_4$), and cerussite (lead II carbonate, $PbCO_3$), are of much less importance. World consumption of lead now amounts to almost four million tonnes per year, of which the UK manufactures 340 000 tonnes (WBMS, 1980).

The metal itself does not occur naturally and is extracted from its ores using long established but ever improving technologies. Crude ore is first roasted in large batches until part of it has been converted to lead oxide and lead sulphate. The air supply is then turned off and the temperature

2

increased. Metallic lead is then extracted with the release of sulphur dioxide. Lead can also be extracted by heating galena with scrap iron. The metal can be further refined using the electrolytic Betts process. Today silver is produced as a byproduct of lead manufacture, being largely obtained from lead ores by the Parkes process, where it is partitioned into molten zinc and extracted by distillation.

The metal has many uses and its malleability allows it to be made into sheets where it is utilized in roofing and plumbing, and for sheathing cables. One third of all industrial lead production is used in the manufacture of electric storage batteries. It is employed industrially in plant for the production of sulphuric acid and for making bullets, shot and weights. Its high density allows its application as screening against radiation around nuclear reactors, X-ray generators or in the storage and transportation of highly radioactive materials.

Its alloys include solders, typemetal, pewter and various bearing metals. The compounds of lead also receive extensive industrial and commercial application, and lead was used in the manufacture of domestic paints as a pigment and dryer. Coloured lead pigments included white lead, red lead, chrome yellow, chrome red, chrome green and brunswick green.

Lead monoxide, litharge, PbO, is used in the manufacture of lead glass and as a dryer in paints and varnishes, where it catalyses the process of atmospheric oxidation and hardening.

Lead dioxide, PbO_2, is a powerful oxidizing agent, once used in the manufacture of matches, but now principally in accumulators.

Trilead tetroxide, red lead, minium, Pb_3O_4, is mixed with linseed oil and used principally as a paint to prevent the rusting of steel, particularly to objects exposed to the weather, such as ships, bridges and agricultural machinery. It is also used for making lead glass and special lead jointing compounds.

Lead II carbonate, white lead, $PbCO_3$ has been used as a pigment for over 2000 years. It is usually made by a modification of the 'Dutch' process whereby sheets of lead undergo slow decomposition under the action of acetic vapour, moist air and carbon dioxide. It is very toxic when used in paints and in recent years has largely been replaced as a pigment by titanium dioxide, on account of its greater opacity, covering power, cheapness, stability and of course, safety.

Lead II chromate, chrome yellow, $PbCrO_4$ is also a pigment, as is the basic chromate, $PbCrO_4 . Pb(OH)_2$, which is known as chrome red.

Other divalent plumbous compounds are used in more specialized applications: lead II nitrate, $Pb(NO_3)_2$, and lead II chloride, $PbCl_2$, are used as oxidizing agents (fluxes) in plumbing, soldering and tinning of metals. Lead II acetate, $(CH_3COO)_2Pb$, and basic lead II acetate, $(CH_3COO)_2Pb . Pb(OH)_2$ are both used in the modern production of white lead.

Finally, lead silicates, silicoborates, and silicofluoroborates are used in the

3

production of many varieties of glass, are constituents of crystal used in optical instruments, and in the various glazes and enamel colours used in the pottery industries.

With the exception of the nitrate and the acetate all plumbous compounds are insoluble in water.

Most tetravalent plumbic compounds are unimportant because they decompose readily on heating and are hydrolysed to lead dioxide, even by water. The exceptions to this are the organolead compounds, e.g. tetraethyl (TEL) and tetramethyl (TML) lead used as antiknock agents in petrol. They are usually added to petrol in a varying composition of tetraethyl, triethyl-methyl, diethyldimethyl, ethyltrimethyl and tetramethyl lead in proportions designed to give maximum overall efficiency (Waldron and Stofen, 1974) and act as a negative catalyst during the combustion of petrol vapour, moderating its explosive violence.

Unlike the diesel engine, in which an even mixture of fuel spray and air ignites spontaneously, the petrol engine requires ignition, via the spark plug. This is a localized event, and there is a progressive flame front in the premixed fuel and air mixture as it burns. The flame front finally dies out in the corner furthest away from the spark plug, and the fuel in this corner (the end gas) is subject to heating and compression for longer, especially when the engine is working hard. Under such conditions, the end gas becomes partially oxidized and ignites spontaneously. This is engine 'pinking' or acceleration 'knock'. At high or persistent loads it will reduce engine efficiency, eventually causing engine damage and failure. The addition of alkyllead compounds to petrol to increase octane quality was discovered in the 1920s. This makes it possible to use a higher compression ratio, with a consequent gain in efficiency. Virtually any metal may be used as an antiknock ingredient, as the key property for this reaction is the ability to form a diffuse aerosol in the cylinder head prior to combustion. However, most metals used in this fashion cause engine wear and corrosion. Lead is one of only a few metals that does not. Early trials showed that lead oxides were deposited on the walls of the cylinder after combustion, therefore small quantities of a 'scavenger', ethylene dibromide ($C_2H_4Br_2$) were also added to convert these lead compounds into volatile lead bromide, which is swept out of the engine by the exhaust gases into the atmosphere, where it is converted into lead carbonate, oxycarbonate and oxides.

Organolead production is the second most extensive user of lead and is the major source of atmospheric lead pollution, about 98% of contamination coming from this source (Lin Fu, 1979). Associated Octel, the sole producer of alkyllead in the UK, is the largest manufacturer in the world outside the United States. In the UK, 60 000 tonnes of lead are used annually in tetraalkyl lead production, of which over 80% is exported. The remainder is added to petrol for the home market, of which 7000 tonnes are emitted as fine particles from the exhausts of cars. This can be compared with the atmospheric release of about 200–250 tonnes of lead during its smelting and refining (UK DoE, 1974).

It is apparent that lead is a very versatile and useful element and that the metal, alloys and compounds of lead form a part of the fabric of industrial and everyday life.

THE HISTORY OF LEAD

The ancient world

At the ancient Hittite city site of Catal Huyuk in Turkey beads of lead have been uncovered that date back to about 6500 years BC (Gale and Stos-Gale, 1981). Lead has been found in a sixth millennium context at Yarum Tepe in Iraq, at the fifth millennium site of Arpachiyeh in Iraq and at the fourth millennium sites of Anau I in Turkestan, Hissar III in Iraq, and Naqada in Egypt. These finds suggest that lead smelting, albeit on a small scale, began at least as early as nine thousand years ago. In the British Museum in London, the oldest artefact of lead is EA 32138, a lead statuette found at the temple of Osiris on the site of Abydos, and purchased in Egypt in 1899. This object is dated on stylistic grounds to the predynastic period of Egypt (circa 3800 BC), and is shown in Figure 1.

To place these dates in a historical context, the age of metals began in the Mediterranean world some time in the middle of the fourth millennium BC, well before the transition of the relatively unspecialized and essentially agricultural society of the late Neolithic into Phase I of the Early Bronze Age (usually designated from about 3500 BC to about 2900 BC in the Aegean).

It can be seen that the association of man and lead goes back millennia, not centuries. The importance of lead in pre-Roman times however, lay in its association with silver. The production of silver as a by-product of burning galena was probably discovered at an early date. In its bare essentials the winning of silver in this way presented no great difficulty. A piece of galena is reduced to lead when heated in an ordinary wood fire: if the heating is prolonged, this lead is oxidised into a powdery ash (litharge, PbO), leaving a small drop of pure silver. As argentiferrous galena contains variable but generally small amounts of silver, bulk quantities could only be obtained when a number of technical problems were overcome. Success was credited to tribes living on the Black Sea near Pontus in Asia Minor in the early centuries of the third millennium (Forbes, 1950). The critically important discovery of cupellation was made next. In this process the lead oxide produced by burning ore is absorbed onto a large bulk of such material as bone ash. This process was probably in use around 2500 BC, since silver objects of high purity had appeared by that date. As a result of these technical discoveries, lead smelting industries arose in several places in the ancient world, particularly in Asia Minor (Aitchinson, 1960).

Although lead was not extensively worked by the ancient Egyptians, it was one of the earliest metals known, since its use dates from predynastic times (before 3800 BC). There is evidence to suggest it was used for sinkers in fishing nets, in glazes, glasses and enamels, and for ornaments. The Ptolemaic and later papyri mention the plumber who manufactures and

Figure 1 A statuette of lead date circa 3800 BC. (Reproduced by kind permission of the Trustees of the British Museum.)

repairs water pipes (Lucas and Harris, 1962). At Troy, Gowland (1912) refers to shapeless lumps of lead found in Troy I, the lowest city (3000–2500 BC). A small lead statue of a naked goddess was found in Troy II, the prehistoric fortress (2800–1900 BC). Lead was used in Crete as early as 2600–2400 BC in the form of small votive axes in tombs. It was also used as the lining of stone chests in the palace at Knossos (2100–1900 BC). The Phoenicians believed they could talk to the dead by placing messages written on little rolls of thin sheet lead into tombs. They also made coffins from lead.

Lead was known in the early Sumerian period when it was known as the metal of Ea, the god of Eridu. The association of metals with gods and planets goes back to the Sumerians where lead is Ninmah (the mother goddess) probably associated with Ninurta (Saturn). An old Assyrian code of laws of 2000 BC shows that lead was used as currency, and later (1400–1050 BC) lead, in the form of animal heads, was a common form of exchange. Lead tumblers were found in very early graves below the royal cemetery at Ur. Six rolls of lead with Hittite inscriptions of the 9th–7th centuries BC were found at Assur. An inscription on lead was found at Nineveh, and fragments of thin sheets on which amuletic texts were written were found at Babylon. Heroditus describes the use of lead in the bridge across the Euphrates at Babylon (*The Histories*, i, 186), and Didorus Siculus mentions the use of lead as a damp course in the walls and on the floors of the hanging gardens of Babylon (*The Library of History*, ii, 10). For a discussion of ancient sources see Partington (1934).

The ancient Hebrews were an essentially pastoral and agricultural people, although it is certain they obtained metals, including lead. Ophereth was the Hebrew word for lead, and it is mentioned several times in the Old Testament. Lead is used metaphorically on a number of occasions e.g. in Exodus where it was noted that the Pharoahs host 'sank as lead' when the Red Sea closed.

'Thou didst blow with thy blast, the sea covered them. They sunk like lead in the swelling waves.' (*Exodus*, 15, 10).

A working knowledge of cupellation, smelting and alloying is evident when they are alluded to in the allegorical purification or strengthening of the Israelites by the Lord.

'Once again I will act against you to refine away your base metal as with potash* and purge away all your impurities.' (*Isaiah*, 1, 25).

'The bellows puff and blow, the furnace glows; in vain does the refiner smelt the ore, lead, copper and iron are not separated out.' (*Jeremiah*, 6, 29–30).

*Potash is translated from the hebrew 'bedil' but may also translate as tin or more likely in this case as litharge.

'Man, to me all Israelites are an alloy, their silver alloyed with copper, tin, iron and lead.' (*Ezekiel*, 22, 17–19).

'This is what the Lord showed me. There was a man standing by a wall with a plumb line in his hand. The Lord said to me "What do you see, Amos?" "A plumb line", I answered and the Lord said "I am setting a plumb line to the heart of my people Israel."' (*Amos*, 7, 7–8).

Lead is also mentioned on lists of metals known to the Hebrews.

'Anything which will stand fire, whether gold, silver, copper, iron, tin or lead you shall pass it through fire and then it will be clean.' (*Numbers*, 31, 22–23).

'Tarshish* was a source of your commerce, from its abundant resources offering silver, iron, tin and lead as your staple wares.' (*Ezekiel*, 27, 12).

Lastly, the Hebrews also knew how to use lead.

'O that they might be engraved in an inscription cut with an iron tool and filled with lead to be a witness in hard rock.' (*Job*, 19, 24).

'Then a round slab of lead was lifted and a woman was sitting there inside the barrel he said "this is wickedness" and he thrust her down the barrel and rammed the leaden weight upon its mouth.' (*Zachariah*, 5, 7–8).

Lead was certainly one of the first metals to be mined by man in the known major civilizations. Lead mines were worked in Sardinia, Athens and Carthage. The Phoenicians mined it in Spain around 2000 BC and these workings were taken over by Rome after the fall of Carthage. Some of the compounds of lead, such as white and red lead, are amongst the oldest pigments known, and it is possible to trace their history back 2500 years (Xenophon, 400 BC). White lead was used as a pottery glaze, and other lead compounds have been used as cosmetics. Some metallic compounds, such as galena and malachite (native basic copper carbonate) may have been used as eye-paints in Egypt from the prehistoric Baderian period, about 5000 years BC (Lucas and Harris, 1962). The practice of using galena for eye-paint survives in fact to the present day, particularly in India where it is known as surma and in Asian immigrants (Attenburrow *et al.*, 1980).

Evidence is available to suggest that the use of lead in the ancient world was not restricted to the emerging cultures close to the Mediterranean or Middle East. In India, lead was used for making amulets and there is evidence of its use in China where it was used as a stimulant in the court of the emperor. It has also been found in pre-Columbian Mexico, where a lead amulet has been discovered from that period (Waldron and Stofen, 1974).

*Lead was amongst the metals traded by the Phoenicians of Tarshish (probably in Spain) who obtained it from the Kassiterides (usually believed to be the British Isles) and from the Andalusian mines in Spain (Partington, 1934). All biblical quotations taken from the *New English Bible*, Oxford University Press, Oxford, 1970.

Although its importance initially lay in its close association with silver, lead emerged from the background and assumed a dominant role in the technology of the developing Roman Empire. Amongst other reasons, the Roman invasion of Britain in the first century was launched to exploit the lead (and tin, copper and silver) mines of England, to satisfy the Roman enthusiasm for sanitation and bathing. The latin word for lead was plumbum, denoting water conduits or spouts; it is derived from the word plumber.

Lead-lined pots were used extensively in cooking, as they prevented the bitterness caused by using bronze containers, imparting a sweet flavour to food. Wine was also prepared in lead-lined containers, specifically because of this sweetening property. The ability of lead to inhibit enzyme activity was well appreciated, and sapa, or grape syrup boiled in leaden vessels, was used extensively as a preservative for fruit and wine. Both Cato (*De re rustica*, cv) and Pliny (*Historia naturalis*, xiv, 21; translation of Jones and Rackham, 1938–1963) advocated the treatment of wine in leaden vessels. These practices caused considerable contamination of food and drink, and Gilfallen (1966) has proposed the doubtless excessive idea that the fall of Rome was due to endemic lead poisoning.

Together with reports of the use of lead are descriptions of its toxic side-effects. These were known to ancient physicians, and the first report is ascribed to Hippocrates (370 BC) who noted symptoms of a metal colic in a metal worker (Jones and Withington, 1923–1931). However, there is no reason to believe that this colic was due to lead, so that this supposition is based more on recent tradition than fact (for a discussion of the issues involved, see Waldron, 1973). The first accurate account of lead poisoning is probably that in the *Therica and Alexipharmaca* (i, 600) of Nicander (2nd century BC), who described symptoms of poisoning by ceruse (white lead) as constriction of the palate and gums, asperity of the tongue, hiccups, a dry cough, nausea, heaviness of the head, unnatural vision and torpor (Major, 1965).

Of the Romans, Vitruvius (1st century BC) mentions that water impregnated with lead was injurious (*On Architecture*, viii, 3), and noted the pallid appearance of lead workers. Horace (1st century BC) is another of the writers of antiquity to mention the purity of water in relation to lead pipes. In a letter to an old friend, he extolled the virtues of living in the countryside, comparing amongst other things, the purity of town water to fresh countryside water.

Purior in vicis aqua tendit rumpere plumbum
Quam quae per pronum trepidat cum murmure rivum?
(*Horace Epist.*, I, 10, 20–21; translation of Wickham, 1891).

Lead has been used as a remedy for thousands of years. It is mentioned in the Ebers papyrus (1550–1500 BC), an Egyptian medical treatise. Lead is specified for laying on a wound (for cooling?). Some Egyptian medical recipes were given by later authors, including Dioscorides, Pliny and Galen. Litharge was the *'spuma argenti'* of Celsus (1st century BC) who mentioned it as a cooling and cleansing medicine. He also treated ceruse as a poison (*De Medicina*, v, 27, 15) but recommended it for burns and ulcers (*ibid.*, v, 7). Pliny (the elder, 23–79 AD) noticed the deleterious effects of exhalations from lead mines (*Historia Naturalis*, xxxiv, 47, 50; translation of Jones and Rackham, 1938–1963), and Dioscorides (1st century AD), noted the toxic side-effects and some early attempts at industrial hygiene in the Roman ship-building industry (*De Venenis*). In his *De Universa Medicina* is found the first mention of lead (as the acetate) as a remedy. He also recommends washed litharge as a remedy for ophthalmic problems, unseemly scars, wrinkled faces and spots. Dioscorides is also the first source to describe the powerful cooling and refrigerant properties of lead (Goodyer and Gunther, 1934). Galen (138–201 AD) mentions that water conveyed in leaden pipes sometimes causes dysentery (*Med.Sec.Loc.*, vii), and expressly says that ceruse ought not to be administered internally (*Methodus Medendi*, iv). For a fuller explanation of roman sources see Scarborough (1969).

With the growth of the Byzantine empire and the transfer of power to the east, Constantinople became the centre of medical knowledge in Europe. The early Byzantine authors tended to paraphrase their Roman predecessors and Oribasius (325–403 AD) and Aetius (early sixth century) both quote many of the Greek and Roman writers. The greatest Byzantine physician is Paul of Aegina (626–690 AD). His report of an epidemic of colic terminating in paralysis is the earliest known description of the clinical picture of lead poisoning (*De Re Medica*, iii; translation of Adams, 1864–1867). He also had a novel use for lead – 'a plate of lead worn upon the loins restrains libidinous dreams'.

Following the collapse of European culture in the 6th–8th centuries, medical knowledge and practice became fossilized in the hands of the church. It was only in the expanding Moslem world that intellectual inquiry continued. The early Arabian physicians, such as Rhazes (865–925 AD) and Serapion (9th century) tended to be bound by the authority of the ancients, supplying little or no additional information. The use of lead as a remedy however, was expanded. It was recommended generally as an astringent in fetor of the armpits to restrain sweating, and to dispel extravasated blood. It does not appear to be used internally, although Avicenna (980–1037 AD) mentions its usage in fluxes and ulceration of the intestines (*Q'anun*, ii, 2, 460; translation of Gruner, 1930). Ibn Baithar (1197–1248 AD) recommended it for diarrhoea, and reported it as being useful in congenital hernia and other complaints around the scrotum.

Although some of these early reports may contain inaccuracies owing to contamination with other toxic metals (notably arsenic, bismuth and antimony), reported symptoms of colic, palsy and paralysis compare favourably with current descriptions.

10

The Middle Ages

Following the fall of Rome in the fourth century, the use of lead declined in Europe and remained at a low level for about 600 years. After the ninth century lead began to be mined in Eastern Germany (the lead-rich regions of Spain were not accessible as they were in Moslem possession). It is known that white lead was used in England in the thirteenth century as there is reference to its use in the Close rolls of Edward I (1274). The practice of adulterating wine with lead and its salts had become widespread, and was banned by the Papal Bull in 1498. Nevertheless, this continued and epidemics of lead colic were by no means infrequent. These were known as poitou colic, the entrapado of Spain, the huttenkatze of Germany, the bellain of Derbyshire and the dry bellyache of the New World. Periodic outbreaks of what are now seen to be lead colic included Poitou colic (1592), Devonshire colic (1793), West India dry gripes (1786), Jamaica dry bellyache (1786) and Madrid colic, (1796). Poitou colic, or 'colica pictonum' was described by Francois Citois in 1616 who thought it was due to unripe grapes. However, it was long afterwards discovered (Tronchin, 1757) to be lead colic.

In England, Devonshire colic was described by John Huxham (1739), but its true cause was not ascertained. This was left to Sir George Baker (1767) who demonstrated that the cider of Devon contained lead, while that of other areas did not. This was due to the common practice of lining cider presses with lead which subsequently dissolved into the mixture. He was responsible for the abandonment of this practice, and thus for the disappearance of the colic. In France, Tronchin (1757) also discovered many wines were able to dissolve the glaze of storage jars, which were compounded with litharge. He demonstrated that a type of colic known as *'bellon'*, which was associated with such wines, was caused by lead.

In the New World, a law was passed in Massachusetts in 1742 to prevent the distilling of wine in lead heads or pipes (McCord, 1953). This was not heeded further south, and 'West India dry gripe' was first described by Thomas Cadwalder. This was an account of lead poisoning from the habitual consumption of Jamaica rum distilled through leaden pipes (Cadwalder, 1745). In Jamaica itself, John Hunter (1788) described the same 'dry bellyache' in the garrison and indicated the same causes. Even today, lead poisoning can be found in alcoholics who drink 'moonshine' distilled through lead apparatus.

Lead was still being used medically and in the early part of the nineteenth century for its action on the blood. Salts of lead were found to be haemostatic and were used for the treatment of ulcers because of their ability to coagulate albuminous material. Until recently, lead compounds could be found in the British Pharmacopoea, including 'lead and opium solution' (a mixture of Goulard's water and laudanum) and 'Diachylon' plasters, which use lead oxide as a base. However, the 1980 B.P. does not list any lead or lead-containing preparations.

Throughout history, lead was used for water and sewage piping. As was mentioned by Vitruvius and Pliny, this practice was not good public health. In modern times, the continued use of lead in plumbing does not appear to be based on ignorance of its toxicity. In an address to the fourth National Quarantine and Sanitary Convention in 1860, Jacob Bigelow (1786–1879) noted 'But where shall we fly to escape from east winds and dogdays, . . . from lead pipes for water contrived to kill everybody except the animalcules'. Today lead piping is still present in older houses, although in decreasing quantity as these are gradually cleared. In the construction of new buildings, lead has at last been replaced (by copper) about 2000 years after its ill effects were first noted.

These uses and abuses made many physicians aware of the nature of lead poisoning. Grisolle (1836), and Burton (1840) recognized the well known blue line on the gums, now called the Burtonian line. This line is a layer of reduced lead sulphate particles that stain the epithelial cells of the gums. Interestingly, the Burtonian line is not a specific marker of lead poisoning, and it is seen in other conditions. A blue line may also be seen around the anus.

The best description of lead poisoning was made by Tanquerel des Planches (1839) who published his work on 1217 cases of lead poisoning in Paris (translated Dana, 1848). This work is the classic in the field, and his studies were so complete that later investigators have added little to the clinical knowledge of symptoms and signs of the disease, as summarized below from a present day clinical source (Price, 1978). Devergie and Hervy (1838) used some of the postmortem material from this study to show that lead was present in the tissues of individuals dying of lead intoxication as, for example, following chelation therapy. The histological features found at autopsy were also briefly mentioned. These were expanded by later workers (Kussmaul and Maier, 1872; von Monakow, 1880).

Clinical features of lead intoxication in man

Acute poisoning
This usually follows intense short-term exposure, but may represent an exacerbation of a chronic intoxication as, for example, following chelation therapy. The initial symptoms include a metallic taste in the mouth, vomiting, colic and the passage of black stools. Circulatory collapse may occur, and encephalopathy is a well recognized feature. Sequelae include hepatitis, renal failure and anaemia. Residual neurological disorder is frequent.

Chronic poisoning
The features of frank lead poisoning include haemopoietic, gastrointestinal, renal, and neurological involvement.
Haemopoietic: lead inhibits haem production, giving rise to the excretion of D-amino laevulinic acid and coproporphyrin III. Features of anaemia include pallor, a reduction of haemoglobin with the occurrence of punctate basophilia in erythrocytes and an increase in reticulocyte count.

12

Gastrointestinal: the main symptom is colic. In some cases it may be so severe as to be mistaken for an acute abdominal emergency.

Renal: renal tubular damage may produce a Fanconi syndrome (with aminoaciduria and glycosuria), and hypertension has been described as a sequel of lead-induced renal damage.

Peripheral neuropathy: wrist drop, due to radial nerve involvement, is a classic manifestation of chronic lead neurotoxicity. Clinical and experimental studies support the view that lead induces peripheral nerve lesions.

Central nervous system: the manifestations of saturnine encephalopathy include headache, irritability, insomnia, apprehension, confusion, nightmares and fits. High exposure levels (at least 200 μg Pb/100 ml in children and 500 μg Pb/100 ml in adults) are usually found. Recovery from encephalopathy is often incomplete, and residual neurological damage is frequent.

The Industrial Revolution

The huge increase in demand for lead caused by the Industrial Revolution brought about the problem of industrial disease, of which the most widespread was lead poisoning (Legge and Goadby, 1912). Women and children were employed indiscriminately in all lead processes, including the highly dangerous jobs of pottery glazing, smelting of lead ores and manufacture of lead compounds, particularly white lead (Hunter, 1975). In 1883, the first act of Parliament directed against a specific occupational disease, the Factories (Prevention of Lead Poisoning) Act, was passed. This required lead factories to conform to prescribed standards. After 1900, intensive studies of industrial hygiene in the lead trades were carried out by such pioneers as Oliver, Legge and Goadby in Britain, Méillère in France and Hamilton in the USA. As a result, a large body of legislation was passed to safeguard workers and to compensate them for their disabilities.

In Great Britain, the work of the first Medical Inspector of the Factories Inspectorate, Dr T. M. Legge, actively investigated the question of lead poisoning from 1892 onwards, with the result that notification was enjoined by section 29 of the Factory and Workshop Act (1895), which consequently became section 73 of the Act of 1901. This enactment requires every medical practitioner attending on or called to see a patient believed to be suffering from lead poisoning contracted in a factory or workshop, to notify the case forthwith to the Chief Inspector of Factories at the Home Office. A similar obligation was imposed on the owner or manager of a factory or workshop to send written notice of such cases to the local Factories Inspector. Following this, a gratifying fall occurred in the incidence of this disease. Lead encephalopathy virtually disappeared from industry, and it was unusual to find cases of severe colic or extensive palsy. The cases that did arise were few and mild

(Legge and Goadby, 1912). This decline is especially marked when it is remembered that during this period consumption of lead increased steadily.

Present day lead pollution

In the last 60 years, significant numbers of cases of lead poisoning have come from the extensive use of alkylated lead compounds tetramethyl and tetraethyl lead (TML and TEL). From 1923, when TEL was first added to petrol as an antiknock agent, cases of lead poisoning associated with its use began to appear, causing considerable alarm. The cleaning of storage tanks and indiscriminate handling of TEL by workers and chemists caused numbers of deaths, and its manufacture was prohibited in 1925, pending investigation by the US Public Health Service. Awareness of the toxicity of TML and TEL, their rapid absorption across skin and lungs and the establishment of more stringent safety precautions in the manufacturing industries allowed their reintroduction in 1926. Organolead compounds were introduced into Britain against opposition during the 1930s when it was established that their addition to petrol was not a serious health hazard (Kehoe *et al.*, 1934).

Table 2 Maximum lead content of petrol (legal restrictions)

Country	Lead concentration (g Pb/l)	Date
USSR	none	1959
Japan	0.02 g/l	1975
Western Germany	0.15 g/l	1976
Switzerland	0.15 g/l	1978
USA (EPA)	0.13 g/l	1980
Sweden	0.15 g/l	1981
EEC (maximum)	0.40 g/l	1981
EEC (minimum)	0.15 g/l	1981
Australia	none	from 1985
UK	0.15 g/l	from 1985
UK	none	from 1990

Nevertheless, considerable concern has been expressed about the pollutant nature of the exhaust products of leaded petrol combustion and about atmospheric lead levels. Atmospheric lead, derived in part from combustion of lead-treated petrol, is seen as being environmentally unacceptable and a potential health hazard, especially to children. In recent years, legislation has restricted the amount of lead in petrol in many countries. Table 2 shows the maximum lead content of petrol in several countries.

In summary, lead poisoning as a recognizable clinical condition has become rare in recent times, due to the introduction of stringent safety precautions in industry and restrictions elsewhere. It is salutary that the

14

number of deaths attributable to this cause has fallen to a very low figure during the course of this century. However, the use of lead is still increasing and there has been much speculation in recent years as to whether the massive release of lead into the environment consequent to its addition to petrol is having an adverse effect on public health.

2
Lead and Man

THE METABOLISM OF LEAD

Lead is ubiquitous in everyday life – it is in the atmosphere, the soil, and present in varying concentrations in food and drink. When Devergie and Hervy (1838) first suggested that lead was present in the body normally, they initiated a controversy that was not finally resolved until the advent of methods sensitive enough to detect traces of lead with reasonable reliability. Today the presence of the human body burden of lead is unquestioned.

Uptake and excretion

Lead may be taken up into the body by inhalation, ingestion, or absorption through the skin, the latter route only becoming important for the alkyl lead compounds in restricted hazardous environments (Lange and Kunze, 1948).

Intake of lead via inhalation is dependent on atmospheric concentration, particulate size and solubility in tissue fluids. Approximately 30% of inhaled lead is retained by the lungs, and clearance from the lungs occurs when the inhaled material is sequestered by alveolar cells or removed via the lymphatic vessels to local lymph nodes. Some inhaled material (about 5%) is absorbed by the mucosa and passes to the gastro-intestinal tract. In general, however it is thought that the contribution of inhaled lead to body burdens is relatively small (Lawther, 1972).

The dietary input of lead has been assessed at approximately $300 \mu g$ per day, which is excreted in a dynamic equilibrium, (Kehoe *et al.*, 1933). Of this input, 5–10%, with wide individual variation, is absorbed from the digestive tract (Kehoe, 1960), both passively by diffusion, and actively via a cationic pump (Gruden, 1975).

Other factors such as age may also modulate lead absorption. In children intestinal absorption can account for as much as 20–50% of input (Alexander, 1974). Children also tend to retain lead for longer periods than adults (Momcilovic and Kostial, 1974), possibly owing to active bone deposition.

17

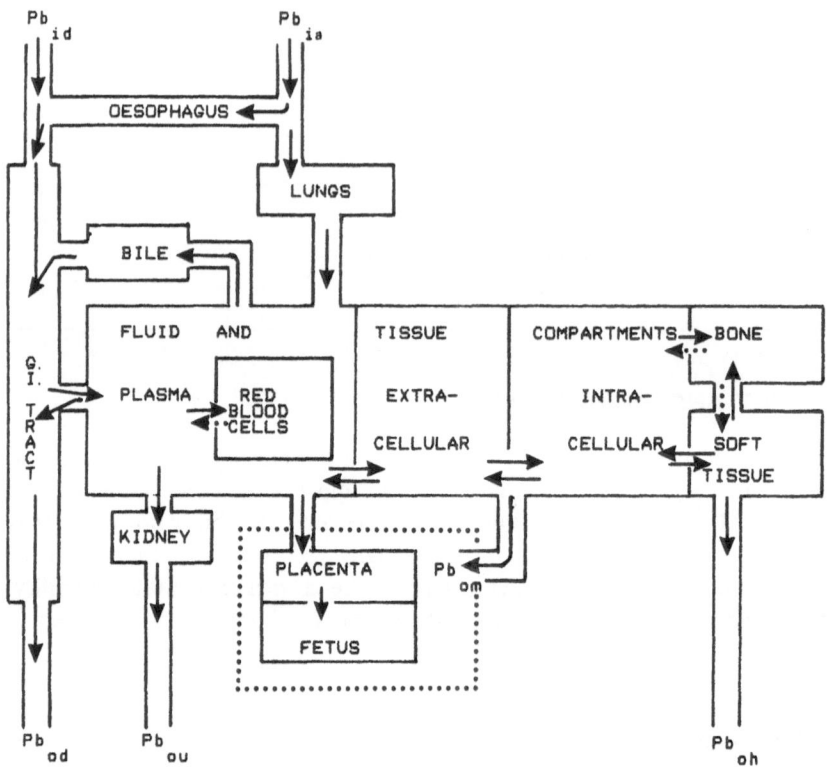

Key to abbreviations and symbols:
  Pb_{id} – lead input via digestion
  Pb_{ia} – lead input via atmosphere
  Pb_{od} – lead output via faeces
  Pb_{ou} – lead output via urine
  Pb_{oh} – lead output via hair, skin, nails, sweat, etc.
  Pb_{om} – lead output via milk (during lactation)
  → main routes
  ⋯▶ minor routes

Figure 2 The bodily metabolism and compartmentation of lead

Diet also plays a large part in determining uptake. Deficiencies in calcium, phosphorus and iron increase the absorption and toxicity of lead (Granick *et al.*, 1978). Other metals such as zinc, aluminium and selenium have protective effects. Others (cadmium and mercury) have synergistic effects. There is considerable metal-lead interaction and the amounts found in the diet mean that nutrition is of great importance for any given degree of lead exposure. Lesser influences on the absorption of lead include previous exposure to lead, where there has been some adaption to toxicity, and alcoholism, which seems to reduce initial lead toxicity without affecting cumulative chronic effects.

Of this uptake, 10% is excreted in the urine, and soft tissue lead is excreted in sweat, hair, nails, skin etc. (Kehoe, 1960). The majority of lead excreted however, is in the faeces, along with unabsorbed lead. The main route of excretion from the body into the faeces appears to be via bile and pancreatic fluid. Cikrt (1972) found that 7% of a dose of radiolabelled lead administered intravenously was excreted into the intestine in bile during the first 24 h. He also showed that lead is excreted through the intestinal mucosa, mainly in the jejunum, ileum and caecum (see Figure 2).

NEUROPATHOLOGY OF LEAD INTOXICATION IN MAN

The most valuable study of the neuropathology of lead encephalopathy in man is that of Blackman (1937), who examined the brains of 22 children with lead poisoning*. He confirmed the occurrence of brain swelling, with flattened convolutions, pressure cones and oedematous white matter. Widespread punctate haemorrhages were sometimes seen macroscopically. Microscopically, numerous lesions of blood vessels, grey and central white matter were evident throughout the brain. Dilated and narrowed capillaries, necrotic vessels, capillary thrombi, minute haemorrhages, and foci of necrosis with reactive glia were seen, as well as perivascular and diffuse oedema. Blackman believed that the parenchymal lesions observed were secondary to damage to small blood vessels. The vascular lesions affected meninges and grey and white matter, and were likened to those seen in acute inflammation. Prominent degenerative cytological changes in arteriolar and capillary walls were observed. Increased vascular permeability was manifested both in low-protein, fibrin-free oedema fluid and in inspissated vesicular eosinophilic material forming droplets up to $50 \mu m$ in diameter clustered about damaged capillaries. Similar material has been noted and examined by other workers (e.g. Smith et al., 1960; Clasen et al., 1974). Coarse vacuolation of grey and white matter was noted by Blackman to occur in various brain regions. The superficial zone of the cerebellar molecular layer was notably affected, with the formation of subpial fluid blebs. In vacuolated areas, damage to axons and myelin sheaths was followed by the appearance of lipid macrophages, both as interstitial foci and around arterioles. Astrocytic proliferation, diffusely throughout the white matter and focally in areas showing spongiosis, is a prominent feature (Blackman, 1937; Okazaki et al., 1963).

Nerve cell changes noted in autopsied cases with lead encephalopathy by several workers (Blackman, 1937; Smith et al., 1960; Okazaki et al., 1963)

*A multiplicity of terms exists to describe lead poisoning. Strictly speaking 'frank' means unmistakable, 'clinical' is recognisable by a physician and 'classic' as described in a textbook. 'Symptomatic' is in the presence of well-defined symptoms and 'asymptomatic' in the presence of an elevated blood lead without obvious symptoms. Other terms include 'subclinical' and 'intoxication' in which effects of lead are assumed but not yet proven.

may be the consequence of focal ischaemia, brain swelling and hypoperfusion rather than primary manifestations of lead neurotoxicity. The existence of direct neuronal effects of high level lead intoxication in man is still uncertain. Clinical anecdotes suggesting the occurrence of primary lesions of motorneurons in chronically lead-exposed subjects are unsupported by autopsy evidence or even long-term follow-up (see e.g. Simpson *et al.*, 1964).

Lead levels at which encephalopathic changes occur in man

Clinically, lead encephalopathy is stated to occur in childhood at blood levels over $100 \mu g$ Pb/100 ml (Betts *et al.*, 1973). However, the nature of the cerebral disorder in documented cases with relatively low blood lead levels in the encephalopathic range is inconstant and reported data suggest multiple pathological processes in some instances. Moncrieff *et al.* (1964) and Betts *et al.* (1973) show that considerably higher levels in children may be without effects on the nervous system (cases with $310 \mu g$ Pb/100 ml and 288 and $380 \mu g$ Pb/100 ml respectively). Higher levels still may be associated with only minimal CNS symptomology in adults (about $1000 \mu g$ Pb/100 ml; Chamberlain and Massey, 1972), while Greengard (1966) reported very high blood lead levels (up to $825 \mu g$ Pb/100 ml) in some children with encephalopathy.

The relationship between the onset of encephalopathy and blood lead level is unlikely to be a simple one, age of subject, acuteness of exposure, and intercurrent illness being only three of the factors accounting for individual variation. However, experimental studies such as those of Goldstein *et al.* (1974), LeFauconnier *et al.* (1980) and Collins *et al.* (1982) suggest that blood and brain levels are roughly comparable with medium to longterm exposure, indicating that blood lead measurements in many cases will be a fair guide to brain load. Fatal encephalopathy has occurred in a child with a blood lead of $336 \mu g$ Pb/100 ml (Betts *et al.*, 1973) and in two adults with brain levels of 580 and $1020 \mu g$ Pb/100 ml (Whitfield *et al.*, 1972).

It is known that the majority of non-fatal encephalopathic cases do not revert to a neurologically normal state, only about 20% of children showing a full recovery (Perlstein and Attala, 1966). This observation is in keeping with the view that lead encephalopathy is due to structural changes in the brain, presumably on a vascular basis as already described. While the clinical data suggest the possibility that such structural changes may occur at blood lead levels about $100 \mu g/100$ ml, levels three times as great may not necessarily produce brain damage, while levels above $800 \mu g$ Pb/100 ml may be attained in children with encephalopathy.

With the decline in interest in the classical aspects of lead poisoning, attention has focussed on the less dramatic consequences of elevated lead burdens in children. Controversy still prevails as to whether lower intensities of exposure can cause subtle changes in CNS function (W.H.O., 1977). There is considerable evidence that neuropsychological and behavioural differences exist between children with low and elevated lead burdens (for reviews, see Rutter, 1980 and Bornschein *et al.*, 1980a), and it is a scientific priority to place these functional alterations in the context of changes in brain morphology. As the possibility of establishing such a correlation in children is obviously untenable, models of lead administration to experimental animals remain the only method of investigation.

THE NEUROCHEMISTRY OF LEAD INTOXICATION IN MAN

Although most neurochemical studies require fresh brain tissue for analysis, it is possible to study aspects of neurotransmitter metabolism with non-invasive techniques applicable to human research. Amine metabolites in urine have been well studied in children and adults. Thus in the course of measuring urinary homovanillic acid (a dopamine metabolite) and vanillyl-mandelic acid (a noradrenaline metabolite) in lead-exposed mice, a parallel study in children was conducted (Silbergeld and Chisholm, 1976). The observed increases in urinary metabolites seen in children were interpreted as suggestive of lead-related changes in brain neurochemistry. However, the multiple sources for urinary HVA and VMA make it impossible to ascribe the observed increases exclusively or even mainly to the effects of lead on the CNS. Furthermore, it is not possible to conclude whether the changes reported in urinary 5-hydroxyindolacetic acid (a 5-hydroxytryptamine metabolite) of lead-exposed workers are reflective of effects of lead on 5-hydroxytryptaminergic function in the CNS (Dugandzic *et al.*, 1973).

The interaction between lead and tetrahydrobiopterin metabolism has also received some attention, and serum biopterin derivative levels have been positively correlated with blood lead levels in human patients (Leeming and Blair, 1980; Blair *et al.*, 1982).

Other neurochemical studies have not been carried out in lead-exposed humans. It is difficult to measure neurochemical functions *in vivo* without recourse to such techniques as cerebrospinal fluid collection. However, the reported effects of lead exposure in rodents on plasma concentrations of prolactin (Govoni *et al.*, 1978) and other pituitary hormones (Petrusz *et al.*, 1979), and tetrahydrobiopterin (Leeming and Blair, 1980; McIntosh *et al.*, 1982) would suggest that similar studies in humans might be considered.

THE TERATOLOGY OF LEAD

The fear that toxic agents in the maternal environment might harm the unborn child is an ancient one that was reinforced by the thalidomide tragedy. Although toxicology and teratology are far more advanced sciences now than 20 years ago, it is still not known what factors with an adverse effect on fetal development may be inhaled, ingested or absorbed by the mother, whether in her smoking or drinking, her medicines, the atmosphere she breathes, or her food.

However, it is clear that a wide variety of external factors are teratogenic (e.g. X-rays, rubella, neuroleptics and other drugs). Numerous clinical observations and animal experiments have also suggested the existence of a much less obvious type of effect of exogenous agents upon the growing fetus. Teratogenic defects are not necessarily structural (Karnofsky, 1965), for alterations of behaviour or functional adaption to the environment may also be seen after intrauterine exposure. The term 'behavioural teratology' was introduced by Werboff and Gottlieb in 1963 as the study of disturbances of brain function not associated with overt malformation of brain, but presumed to be due to subtle structural changes. Twenty years later, the nature and extent of such alterations in man are still far from clear. However, evidence from laboratory experiments on animals shows behavioural teratogenicity to be a real phenomenon and hints at possible mechanisms whereby prenatal exposure to an adverse exogenous influence could alter learning ability, motor co-ordination and other complex brain functions.

It must be emphasized that knowledge in this field is still fragmentary. Even where the effects of a toxic substance on experimental animals appears to be clear cut, the occurrence of parallel changes in human brain function often remains to be convincingly demonstrated. Nevertheless there is little doubt that animal observations give a valid insight into what is happening in man, and provide information on the basis of which important economic, social and therapeutic decisions may have to be taken.

Lead as a teratogen

The toxicity of lead to the developing embryo and fetus has been known for centuries. It is an old observation that lead is an abortifacient, and that female lead workers had a high rate of miscarriage (Legge and Goadby, 1912). In a series of 123 cases studied by Paul (1860) there were more than 73 fetal deaths, and of the 50 live births 35 had died by age 3. Later studies noted that abortion and stillbirth rates of lead workers were as high as 60% (Aub et al., 1926).

It is well documented that lead crosses the placental barrier (Carpenter, 1974, Lauwerys et al., 1978) and umbilical cord blood has been shown to contain lead in concentrations similar to those found in maternal blood (Harris and Holley, 1972). Although the abortifacient capacity of lead may be due to a direct effect on the fetus, pathological effects on the placenta may also play a role. Baker (1960) showed placental necrosis and haemorrhage, and Dawson et al. (1969) showed that lead may severely depress oxidative metabolism in the toxaemic placenta.

Paternal lead toxicity may also influence the development of the fetus (Oliver, 1891; Tredgold, 1947). Van Assen (1958) reported the case of children with lethal malformations born to a mother who had had normal deliveries previously, and to a father subsequently diagnosed as having lead poisoning. After the father was treated appropriately, the mother produced normal children again, suggesting a mutagenic as well as teratogenic property of lead. The findings of Lancranajan *et al.* (1975) are of interest in this context, abnormalities of sperm morphology and motility having been shown. Together with alteration in chromosomal integrity (Verschaeve *et al.*, 1979) these changes suggest a direct action of lead on spermatozoa at levels which seem to produce few other, if any, demonstrable effects.

As to other prenatal effects, lead certainly appears toxic to the fetus, although intrauterine exposure is generally believed to be at a low level. In a study by Barltrop (1969) only about 300 μg of lead was found in the fetus at term. Measurements of lead in amniotic fluid show a concentration of less than 20 ng/100 ml, i.e. below the limits of detection for the analytical method used (Kubasik and Volosin, 1972). However, lead is shown to be present in the fetus at all stages of development, and this may be of greater importance than the presence of a given load at term. The distribution of radiolabelled lead in embryonic animals indicates that all major organ systems are exposed (Carpenter, 1974).

Other embryopathic effects have been shown in animal studies. For obvious reasons experimentation with potential teratogens cannot be justified and carried out on humans, so cross-species comparisons must be evaluated carefully. Of the animal studies reported, some show teratogenicity. Anomalies of the tail and sacrum of fetal hamsters have been induced by feeding the mother lead (Ferm and Carpenter, 1967). McClain and Becker (1975) caused a urorectocaudal syndrome of malformation with intravenous injection of inorganic lead, whilst their organolead studies show only nonspecific growth retardation in the rat (McClain and Becker, 1972). This decrease in weight gain has been seen by other workers (Schroeder and Mitchener, 1971; Michaelson, 1973) and it could be due to an effect secondary to maternal toxicity or debilitation. It has not been possible to produce developmental abnormalities either in sheep (James *et al.*, 1966) or cows (Schupe *et al.*, 1967), and although there is considerable circumstantial evidence, there is still no certainty that lead is teratogenic in humans.

CHILDHOOD LEAD EXPOSURE

A blood lead level at or above a defined limit (80 μg Pb/100 ml in 1958, and 30 μg Pb/100 ml in 1982) is often found in children living in poor urban areas. Amongst contributory factors are domestic paint containing white lead (now banned), lead plumbing in old housing and traffic exhaust fumes. The causes and incidence of high lead levels, which raise the possibility of widespread toxicity, have been extensively studied.

Screening programmes

Early epidemiological surveys (Bradley *et al.*, 1956; Griggs *et al.*, 1964) established that lead intoxication was more prevalent than was generally appreciated, and that children were especially at risk. This was due partly to the exploratory activity of mouthing objects and partly to pica, or geophagia, the pathological consumption of non-food objects (e.g. plaster and paint flakes). Pica is a habit frequently intensified by social and emotional deprivation, and by poor nutrition. Pica was early established as a source of lead poisoning in children (Greenberg *et al.*, 1958), but is in fact a relatively rare behavioural disorder (Forfar and Arneil, 1973).

The late sixties saw the creation of large scale screening programmes in American cities. Improvements in atomic absorption spectrophotometry (AAS) made such surveys possible by virtue of rapid accurate lead measurement in large numbers of biological samples (Jacobziner, 1966). These studies showed that up to 40% of children tested in inner city slums have 'elevated' blood lead levels at between 40 µg and 60 µg/100 ml blood. Overt signs or symptoms of poisoning were seldom encountered at these levels.

In 1970 the definition of 'normal blood lead' was clarified when a statement on the medical aspects of childhood lead poisoning was published by the US Surgeon General. This document stated that 'excessive absorption of lead' occurred above 40 µg Pb/100 ml blood as measured on two separate occasions, and that 'unequivocal cases of lead poisoning' occurred above 80 µg Pb/100 ml. Any individual with such a blood lead was to be immediately hospitalized (HSMHA, 1971). These values were used as the criteria for all subsequent studies until the most recent recommendations by the Center for Disease Control were accepted in 1978. The blood lead values giving cause for concern were now reduced to 30 µg/100 ml for 'elevated blood lead'; and 'two successive blood leads equal or greater than 70 µg/100 ml with or without symptoms' (USCDC, 1978). Currently even these values are under discussion for further revision by the EPA in America, the EEC in Europe, and the WHO.

In the early 1970s several studies revealed that excessive lead absorption was not confined to children in urban slums, although these were at high risk. High lead intake was also found nationwide, outside urban slums, in poor housing even in rural areas (Fine *et al.*, 1972; Fine and Dobin, 1975). Estimates of numbers of children with elevated blood lead levels based on the concept that excessive exposure was only found in metropolitan areas were therefore incorrect. However, the results of later programmes (Lin Fu, 1973;

Sachs, 1974) indicated that the incidence of high lead exposure appeared to be falling. This was due to such factors as the changes in the population of the children being screened, increased awareness about lead neurotoxicity, and slum clearance programmes. Death from lead poisoning is now almost zero and encephalopathy is rare. This in turn has focussed attention on the subclinical effects of lead toxicity.

Neurological sequelae

The association between frank lead poisoning in children and permanent neurological damage was known from the early 1930s. Studies of neurological and psychological sequelae subsequent to lead poisoning dealt with children who often exhibited CNS symptoms initially.

Symptomatic

Byers (1959) reviewed the outcome of 45 children previously hospitalized with lead poisoning and found one third to be mentally deficient. Perlstein and Attala (1966) reported that 39% of 425 children with lead poisoning showed neurological impairment in follow-up examinations. This was most marked in those infants whose symptoms included encephalopathy (i.e. 81% of 59). In a review of the literature by Wiener (1970) a link between some degree of mental impairment and lead poisoning was established.

Asymptomatic

It has been suggested that asymptomatic children with elevated lead levels may also show some form of psychological or neurological disturbance. Peuschel *et al.* (1972) followed a series of children with elevated blood lead measurements ($> 50 \mu$g Pb/100 ml). Minor neurobiological dysfunction was believed to be present in about a quarter. Initial psychological assessment revealed low average mental capabilities in the majority of this group, although there was a later improvement in IQ.

Kotok (1972) found that the developmental deficiencies in a group of asymptomatic children with elevated blood lead levels were identical to those of a control group similar in age, sex, environment, neonatal condition and, importantly, incidence of pica. Moreover, these deficiencies could be correlated with inadequacies in the environment. Because a somewhat insensitive measure of development was used (the Denver Developmental Screening Test), and the groups were small, this study could not exclude the possibility of minimal neurological damage. Its main conclusion was that the observed developmental deficiencies could not be considered to be due to lead toxicity.

The validity of Kotok's conclusion, i.e. that pica, rather than lead, is the variable associated with mental retardation has been questioned (see de la Burde, 1972; Chisholm, 1972). A later study by this group also controlled for the presence of pica: no significant differences in cognitive function were found (Kotok et al., 1977).

Other studies by de la Burde and Choate (1975) showed that asymptomatic children with 'high' lead levels scored poorly in cognitive and psycholinguistic tests. Rummo et al. (1979), using 'encephalopathic', 'high lead', 'low lead' and control groups of children, found global neurological and psychological impairment only in the 'encephalopathic' group. Needleman et al. (1979) in a comprehensive and much-cited study of psychological and classroom performance, using dentine lead as a measure of exposure, described many differences between high and low dentine lead groups. These include deficits in intelligence and verbal skills; in auditory or speech processing; and on measures of attention. Another study in Germany, also using dentine lead, reported similar findings (Winneke et al., 1981), and a pilot study in the UK (Yule et al., 1981) again obtained similar results. In the last, 106 children were grouped by blood lead (range 7–32 μg Pb/100 ml) and assessed on a battery of psychometric tests. There were significant negative correlations between blood lead and attainment scores on reading, spelling and intelligence; these differences remained after social class was partialled out.

However, a criticism of these studies, which can be applied to most similar papers, is that an inference of causation, i.e. that excess lead caused the documented deficits, may be unjustified. A more vigorously controlled study by Baloh et al. (1973) merely showed that a higher proportion of children in a high lead group (44% compared with 15%) were considered overactive. Concern was expressed by these authors about the difficulty of controlling or measuring such variables as, for instance, lead absorption during infancy. Problems of overcorrection for control variables may also occur.

As a result of the extensive screening programmes in Chicago of the early 1970s, a cohort of children who had at least one high blood lead were studied longitudinally, (Sachs et al., 1970). In the initial study, the IQ of children whose blood lead had been above 50 μg Pb/100 ml (range 50–365 μg Pb/100 ml) was assessed. Sachs et al. (1978) noted that full scale IQs of lead-poisoned and control groups (below 40 μg Pb/100 ml) did not differ. In a follow up study using the same groups, Krall et al. (1980) also reported that there was no significant difference with scores on tests of perceptual and verbal patterns. In the most recent report from this group, a large battery of neuropsychological tests were given to 100 children who had been intoxicated by lead during childhood. No differences were found (Sachs et al., 1982). It was concluded that no impairment can be characterized as identifying asymptomatic lead exposure above blood lead of 50 μg Pb/100 ml (Sachs et al., 1982).

Ernhart *et al.* (1981) related scores for various child ability tests to indicators of lead exposure. Some of the scores, including those for cognitive and verbal scales and reading tests, were significantly impaired in children in the higher lead groups. However, variance associated with lead decreased when a measure of parental IQ was incorporated into the analyses. In view of this, these authors speculated that it is quite possible that the inclusion of more sensitive measures of parental intelligence or measures of other factors affecting child development may have reduced the variances still further.

Some of the questions posed by Lin Fu in 1973 remain unanswered today. These include (a) the suitability of blood lead as an indicator of exposure, (b) normal blood levels and the possibility of a lead threshold, i.e. a safe level of lead exposure and absorption, (c) the effects of low level lead and, if any, their clinical manifestations and (d) the distinction between lead absorption in symptomatic and asymptomatic individuals. The first of these questions has been partially answered. Although blood lead measurements reflect a compromise between accuracy and convenience, despite obvious limitations, they still remain a reliable means of assessing recent absorption. The other questions remain problematic.

Many papers have attempted to define a normal blood lead level and seem to imply that values not diagnostic of lead poisoning are normal. Indeed, most papers equate the lowest symptomatic blood lead level with the upper limit of normal. A level not associated with overt clinical evidence of toxicity is not necessarily normal. However, most children with increased and therefore 'abnormal' blood leads are reported as asymptomatic. This may be due to the absence of classical symptoms and the failure to recognize other less obvious clinical features. The evidence supporting low lead toxicity is nevertheless slight and is prone to criticism, as other variables, e.g. genetic, perinatal, nutritional and socioeconomic factors may be just as relevant in causing the sequelae detected.

The measurement of IQ is sufficiently imprecise in its correlates with real adaptive behaviour, and studies of lead-induced behavioural impairment should use many other supplementary measures, rather than relying exclusively on IQ. It is necessary to turn to other approaches in the full assessment of lead effects on neurological development, and retrospective studies and epidemiological surveys of environmental hazards are examples which will be considered next.

Retrospective studies

From the early American epidemiological surveys it became evident that many lead-exposed children were remaining undiagnosed because of their non-specific symptoms, and because of a lack of a convenient method for lead determination at that time (McLaughlin, 1956). An alternative approach was attempted, that of retrospective studies, in which children who showed some form of mental disorder had their blood lead content analysed.

The earliest of these was that of Moncrieff *et al.* (1964) which is noteworthy for taking 36 μg Pb/100 ml blood as the upper limit of normal. This level has only become accepted in recent years, having been regarded as too low by earlier authors (e.g. Chisholm, 1965). Moncrieff found raised blood levels in 55% of 122 mentally retarded children but believed that the high lead levels were secondary, inasmuch as retarded children may have marked pica, and are more likely to ingest more lead than normal children. This extra lead ingestion might of course produce further mental impairment (Moncrieff *et al.*, 1967).

Gordon *et al.* (1967) and Gibson *et al.* (1967) also determined blood leads of children with mental retardation of unknown aetiology. In both studies a large proportion of children had blood leads greater than 40 μg/100 ml blood, but again in neither study was the finding considered to be of primary significance. Gibson and her colleagues speculated that modest elevations in blood lead concentration might be associated with biochemical abnormalities in the brain.

A number of children with hyperactivity, for which no conventional cause could be demonstrated, were found to have higher blood leads than control groups. These control groups consisted of a group of normal children and a group of hyperactive children in whom a predisposing cause had been diagnosed (David *et al.*, 1972; David, 1974). Hyperactivity is one of many terms used to describe a complex multifaceted syndrome of disruptive and impulsive behaviour and distractability (Denkla and Heilman, 1979). This resembles the disorder also now known as minimal brain dysfunction (MBD), a term sometimes regarded as a residual diagnosis category. Although MBD has been considered to be a genetic disorder it is also alleged to involve disturbed monoamine metabolism (Wender, 1978). It is conceivably relevant that animal studies suggest a link between lead, monoamines and hyperactive behaviour (Golter and Michaelson, 1975); this connection will be discussed in detail below. In the much quoted studies noted above, Needleman *et al.* (1979), Winneke *et al.* (1981) and Yule *et al.* (1981) correlated hyperactivity in children with lead levels in teeth or blood. David *et al.* (1976) have alleged that hyperactivity in children can be successfully treated with lead chelating agents, so that the association between lead intoxication and hyperactivity may not be entirely fanciful.

In a recent report (David *et al.*, 1982), blood lead concentrations were measured in children with mental retardation of unknown aetiology, and a negative correlation between lead levels and IQ was noted. As the mean lead level of mentally retarded children was 25 μg Pb/100 ml and the control group 19 μg Pb/100 ml, this correlation is not easy to interpret. The authors point out that the normal sampling method of individuals in blood lead studies does not take into account biological variability, and only in studies where the sample population contains affected individuals (as is the case in this study) will a true link between lead and brain function be established. The use of thresholds in the diagnosis of exposure is erroneous for the same reason (David, 1980).

In 1975 Beattie et al. reported a controlled study of mentally retarded children in Glasgow, a city that has a high proportion of lead water pipes in much of its older housing. This fact, combined with the softness and acidity (i.e. plumbosolvency) of Glasgow water, resulted in the tap water having a higher lead content than elsewhere in the UK (Hansard, 1976). Beattie argued that there was a suggestion that lead exposure caused mental retardation, as there was a close correlation between the lead content of drinking water and the mental capacity of infants that drank it. In a follow up study by Moore et al. (1977) this correlation was extrapolated back to perinatal life. Analysis of stored blood samples taken at birth for PKU screening showed a highly significant link between maternal drinking water lead content and the incidence of mental retardation in offspring. Moore suggested that intrauterine lead exposure might therefore be important.

Hair lead has also been used to determine exposure. A group of 31 children with learning disability were found to have significantly elevated concentrations of lead and cadmium in hair (Pihl and Parkes, 1977). However, no differences were reported in hair mineral content (including lead) or nutrient uptake in autistic children (Shearer et al., 1982).

While these studies suggest that behavioural or intellectual function could be impaired by exposure to lead, doubt is thrown on such an interpretation by poor or inadequate controls, or poorly defined behavioural parameters. More work both on experimental animals and with more rigorous psychological analysis (e.g. as outlined by Barocas and Weiss, 1974) clearly needs to be done.

Epidemiological studies of environmental hazards

Studies of environmental exposure to lead have concentrated on three main sources. The first group, already discussed, has examined individuals exposed to lead-laden water, lead piping and soft water being causal factors (cf. Thomas et al., 1979). The second and largest group of studies have examined populations living near lead smelters and factories (Elwood et al., 1977; Martin et al., 1975; Roels et al., 1978). The last group consists of studies of possible or actual accidental exposure. Examples of the former are based on populations living close to spoil tips or near lead mines (Thomas et al., 1979) or residents near heavy traffic (Daines et al., 1972). There have also been cases of toxic exposure to lead from the burning of old batteries. An example of this is the Rotherham incident, where high levels of lead were found in children playing in the ashes of burnt accumulator cases.

Most of these studies have shown that blood lead values are invariably increased by exposure, but little or no health hazard has been adduced. This may be due to the fact that minor effects (e.g. behavioural ones), have not usually been investigated. Lansdowne et al. (1974) linked proximity to a lead smelter in London with increase in blood lead levels. There was however no link between lead and any measure of mental capability that was investigated, and any observed differences that were found could be related to social factors. Hebel et al. (1976) compared distance from a Birmingham battery factory with 11-plus examination scores and found that children who lived near the factory since birth scored more than those with a low lead-load or living in unpolluted areas. They also reported that this difference remained when adjustments were made for social class, birth rank and maternal age.

Landrigan et al. (1975) studied groups of children near a smelter in El Paso, Texas. In double blind assessment of various neuropsychological variables of matched groups of children, with high (40–80 μg Pb/100 ml blood) and control (up to 40 μg Pb/100 ml blood) lead levels, differences were found in IQ, verbal, behavioural and hyperactivity ratings which were unchanged when adjusted for age and sex. Landrigan was careful to point out that his group matching procedure was not wholly satisfactory, but he did conclude that blood levels over 50 μg Pb/100 ml might result in minor yet statistically significant impairments of fine motor skills and intelligence. After the installation of pollution controls, a follow up study (Morse et al., 1979) showed that blood lead levels had decreased significantly, but no measurements were carried out to assess any improvement in the previously observed behavioural differences.

Another study by Ratcliffe (1977) of children around a Manchester battery factory showed no significant differences on development and behavioural scores between groups with above and below 35 μg Pb/100 ml blood. However, the small size of the group in this study (lead, $n = 23$; control, $n = 29$) might suggest an insufficient number of subjects.

Mackie et al. (1977) were unable to demonstrate a correlation between tooth lead levels and residence near a potential industrial source. House dust was implicated as a major source of lead by Charney et al. (1980), and Walter et al. (1980) were able to show that household dustiness was significantly related to blood lead in children under 2 years. Soil lead may also be an important source of ingested lead between 2 and 7 years. There was also a slight effect of the occupational category of the fathers. It can be seen that the cause of moderate blood lead elevation is likely to be multifactoral: no single source can account for elevated levels in all children. Milar et al. (1980) noted that the intellectual deficits previously attributed to lead may be related to a compromised home environment.

Changes in the electroencephalogram (EEG) in children with high blood leads have been shown to exist. EEGs of the children whose psychological and classroom performance were studied by Needleman *et al.* (1979) were measured. Differences were seen in spectral analysis of EEG data between high and low lead groups (Burchfiel *et al.* 1980). In another study the EEG of children with blood lead levels above 30 μg Pb/100 ml were said to be altered (Otto *et al.*, 1981; Benignus *et al.*, 1981). However, the functional implications of these changes are unclear.

It can be seen that it is difficult to make firm conclusions from studies of this nature. The methodological assumptions and control or matching procedures must be viewed with caution. It would appear that assessing minor differences in relatively coarse measures, such as intelligence quotient, educational attainment and behavioural patterns, is unlikely to produce conclusive results. It appears that there is no easy way to ascertain whether lead, at the levels of exposure studied, causes mental handicap, or minor intellectual impairment, or indeed any neurological disturbance at all.

SUMMARY

The observed pattern of lead poisoning has changed in recent decades. The dramatic picture of a comatose child with lead encephalopathy has become a rarity. Instead, the possibility has been raised of a syndrome of lead poisoning where minor neuropsychological effects are manifest.

It seems reasonable to infer that raised levels of lead, not necessarily in the toxic range, but present over a long period of time, could conceivably be responsible for the hyperactivity and MBD, mental retardation, intellectual and psychological impairment, and some other behavioural changes already mentioned. Some reports show clear strong associations and others report equivocal, even negative associations. This discrepancy can be traced to a number of procedural deficiencies, i.e. insensitive outcome measures, ascertainment bias, unreliable exposure markers, other confounding variables and strategic differences in study design.

It is evident that concern for the effects of environmental lead pollution is well considered, and it is essential for the medical and scientific communities to study further the problem of low lead toxicity and its effects.

3
Experimental Models of
Lead Administration

The fact that lead has harmful consequences upon the developing nervous system is no longer in doubt, as has been described in the previous chapters. Since the early 1940s there have been many publications describing its devastating effects at high levels. In the last 10 years however, there has been a shift in emphasis in the medical and scientific literature, on the supposition that more subtle effects at lower concentrations can be inferred from the gross changes observed at high levels. However, the evidence for this is often speculative and usually contradictory.

DEVELOPMENT OF THE CONCEPT OF SUBCLINICAL
LEAD POISONING

The clinical picture of overt lead poisoning in young children has long been recognized. Detailed information has been published and levels of exposure at which this may occur are well known. From these, laboratory guidelines have been produced to indicate whether excessive exposure has occurred. Periodically revised downwards, these recommendations use empirically established criteria to define exposure. Although dentine and hair lead, and urinary porphyrins have been used, blood lead, normally expressed in $\mu g/100$ ml blood, remains the most common and convenient measure of exposure.

It is difficult to define a normal blood lead, and as already noted most papers equate the highest blood lead in non-exposed populations with the upper limit of normal. In the USA, abnormal levels of lead exposure are currently defined as $30\,\mu g$ Pb/100 ml (excessive absorption); and $70\,\mu g$ Pb/100 ml (unequivocal cases of lead poisoning), (USCDC, 1978). For Europe, a recent EEC directive sets out guidelines for blood lead levels that should not exceed: $20\,\mu g$ Pb/100 ml for 50% of the group, $30\,\mu g$ Pb/100 ml for 90% and $35\,\mu g$ Pb/100 ml for 98% of the group (CEC, 1980). The issue of such recommendations by national and international agencies is not without its critics (see e.g. Elwood, 1982), for stringent directives are often unworkable in practice. Moreover, it is evident that some children may have blood levels above these limits yet may not show any signs of toxicity, i.e. they are asymptomatic. The nebulous term 'subclinical' has been introduced to describe such individuals (Waldron and Stofen, 1974), on the assumption that the lack of any observed signs of toxicity was based on their undetectability, not their absence. However, it is fair to say that any acceptable level of exposure does not represent a boundary between safe and unsafe, but is merely an arbitrarily chosen point on a continuum.

Animal experimentation on the developmental consequences of lead has not clarified this issue. A great deal of lead research is based on the reasonable premise that much can be learnt about effects in children by studying exposure in developing animals. However, a cursory glance at the literature shows that much animal work is not relevant to human childhood exposure. While there are numerous model systems and dosing regimes available, there are very few that attempt to address the problems that are pertinent to exposure in children.

Animal models of lead intoxication

Initial interest in lead intoxication as a clinical phenomenon was focussed on encephalopathy and the gross neuropathological changes that can be ascribed to lead exposure; only more recently have the less obvious consequences, like behavioural effects and their possible basis, been studied in depth. Attention has been focussed on children as the group most at risk from lead intoxication. As they appear to be more sensitive than adults with respect to the more obvious manifestations of lead toxicity, the same has been presumed to be true for more subtle effects.

Because the human interest lies in effects in children, experimental lead studies have been of two main types. (1) Specific investigation of lead toxicity and (2) studies of the effects of lead on growth and development. A large proportion of the studies have been of the latter kind, at least by design. The distinction between studies of growth and of development needs clarification, as they represent separate conceptual entities. When assessing the possible effects of lead on development, it must be borne in mind any developing organism will have tissues in the process of development (i.e. undergoing proliferation, migration and differentiation), and those that are fully developed but in the process of growth. Effects of an agent like lead, with influences on a variety of systems can be of two types: those that affect development and therefore result in a delay in reaching maturity or a changed composition in the mature tissue; and those that simply retard growth. In some cases it may be difficult to make this distinction. As often happens with the conceptual subdivision of a problem, experimental reality reveals stages

34

of transition. Nevertheless, such a distinction is important for adequate comprehension of abnormal developmental processes which are specific results of a neurotoxic substance and more general effects of growth suppression.

The experimental study of lead encephalopathy in laboratory animals began early in this century, and substantial literature was in existence before the Second World War. Early workers, using mature animals, were able to produce brain oedema, haemorrhage, and reactive gliosis, but found that even very high doses of lead gave only inconsistent results. The study of the effects of lead on the developing nervous system was also sketchy, as no satisfactory animal models existed before 1966. The observation of Pentschew and Garro (1966) that lead encephalopathy could be induced in neonatal rats by adding lead to the maternal diet, greatly facilitated the study of lead poisoning. The risk of lead intoxication is high in suckling rats, despite the fact that adult rats are highly resistant to lead (Sharding and Oehme, 1973). It has been demonstrated that this method delivers significant amounts of lead to the neonate in the maternal milk (Bornschein *et al.*, 1977), and treatment of lactating animals is an efficient method of producing chronic lead exposure in neonates (Hejtmancik *et al.*, 1982). The Pentschew and Garro (1966) model has since been considered a milestone in the experimental study of lead neurotoxicity for it was felt, for the first time, that constant, characteristic and reproducible results could be obtained by a simple procedure. The pathological changes seen in the brain of affected rats resembled those described in children with lead encephalopathy.

The study of Pentschew *et al.* (1966) stimulated experimental interest in lead neurotoxicity, and adaptations of their method have become widely used. While encephalopathy has now been induced in a variety of species using high lead exposures (Table 3), the rat remains the animal of choice, and the most systematically investigated.

Large doses of lead administered to developing rats cause reproducible neuropathological changes, and it is not necessary for lead to be ingested as in the Pentschew and Garro (1966) model. Similar changes can be evoked using lower doses and long exposures, or high doses for short periods of time. The early high dose experiments did little to characterize the Pentschew and

Table 3 Reports of lead encephalopathy in various animal species

Reference	Animal
Pentschew and Garro, 1966	Rat
Rosenblum and Johnson, 1968	Mouse
Hopkins and Dayan, 1974	Baboon
Takeichi and Nada, 1974	Cat
Bouldin *et al.*, 1975	Guinea Pig
Wells *et al.*, 1976	Calf
Stowe and Vandevelde, 1974	Dog
Reyners *et al.*, 1980	Monkey

Garro model, and some peculiar anomalies arose as a consequence. The morphological abnormalities observed were found to be due to the animals being weaned from a 25–50 ppm Pb milk to a 32 000 ppm Pb diet, a load of over 100 times initial levels (Sauerhoff and Michaelson, 1973). It was not possible to restrict lead consumption to the maternal rat alone as the natural exploratory activity of neonate rats meant that they soon obtained access to the high lead diet. These workers therefore refined this model so that the animals were weaned onto a 25–50 ppm Pb diet, thus removing the huge increase in lead load at weaning (Michaelson and Sauerhoff, 1974a). Neonate rats receiving this regime did not show paraplegia or any gross histopathological changes, although some retardation in weight gain was observed. This model assumes that the caloric content of the milk and diet are the same, and that postweaned animals consume the same amount of diet as milk before weaning. This is not the case. One of the main functions of the intestine in the newborn animal is to absorb intact macromolecules, especially maternal immunoglobulins, from the colostrum (the protein-rich milk produced at birth). The mechanism of uptake is via pinocytosis by 'immature' enterocytes in intestinal villi. A new population of enterocytes, which are not capable of pinocytosis, are produced in the rat from about day 18 (Clarke and Hardy, 1969). This transition is dependent on internal or external triggers, such as the presence of food in the gut (Daniels, 1972). The pinocytotic uptake of the lumen contents will obviously increase the amount of lead taken up by the neonate, giving the lead-dosed developing rat a variable uptake as it ages. Estimates of absorption of lead in rats range from 50% in the neonate (Momcilovic and Kostial, 1974), down to 1% in the adult (Kostial *et al.*, 1971).

The addition of lead to diet is still used as a regimen of lead administration, although it has been largely replaced by the addition of lead to drinking water as the most commonly used method. The reasons for this are that it is quicker to dissolve soluble lead salts in drinking water than to prepare special lead-containing diets, and more convenient to monitor lead exposure, as there is little spillage. As with the administration of lead in the diet, this method relies on equivalent consumption to that of control animals, which may not occur (as for example, with high doses). This has been overcome to some extent by the restricted watering schedule of Zenick and colleagues (1979a; b; 1981), although this method may require more time than the others.

Another common method of exposure is to treat the neonate by daily oral intubations (gastric lavage, gavage), permitting a relatively accurate estimate of daily exposure. This method has several advantages over administration via diet or drink, by avoiding some of the errors resulting from unconsumed diet, spillage, leakage, and the possibility that the offspring may gain access to the lead regime, thus increasing their exposure. Another feature is the absence of any maternal intoxication. There are however, several disadvantages with this method: delivery of a single 'pulse' of lead with a short transit

36

time through the gastrointestinal tract (which could cause a periodicity in blood concentration with peaks in the acute toxic range), daily dosing can cause stress to neonatal animals, and in relation to other methods of administration, this method is labour intensive.

It is obvious that these models do not replicate exposure in suckling children, as contamination of human milk or formulae is usually a minor source of lead intake (Lamm *et al.*, 1973; Dillon *et al.*, 1974). However, they preserve two essential aspects of paediatric lead exposure: the oral route of exposure, and the timing of exposure during critical periods of brain growth and development. The comparability of growth patterns between rat and man is such that the preweaning period in the rat is analogous to the first two years of development in humans (Dobbing, 1968). It should also be noted that most of the early models have been used for the study of postnatal lead exposure. The effects of prenatal exposure (the period in which the majority of CNS structures develop) and currently of some clinical concern, are only now beginning to be evaluated in animals. The relevance of the few studies undertaken (e.g. Brady *et al.*, 1975) is not yet established, although the work of Crofton *et al.* (1980) and McCauley *et al.* (1982) seems to suggest that prenatal exposure is more likely to produce lead-induced effects than postnatal exposure.

Other routes include intraperitoneal and intravenous injection, but these tend to be used only in special experimental designs that disqualify other more general methods, e.g. in investigations using radioactive isotopes of lead. These cause great stress, particularly if given under anaesthesia (intravenous) or after prolonged intraperitoneal injections, and are usually avoided. A further complication is that administration of soluble lead salts by injection causes the formation of insoluble salts such as the phosphate and carbonate in blood (intravenous) or peritoneal cavity (intraperitonal). This produces irritation or other toxic effects (Silbergeld and Goldberg, 1974b).

In recent years, interest in the clinical problems related to low-level lead exposure has directed experimental attention away from high-dose systems with vascular encephalopathic changes to others where lower doses of lead are given in an attempt to identify less dramatic neurotoxic effects. Table 4 lists a number of experimental systems abstracted from the literature of the past 16 years. In these, rats have been subjected to lead exposure via every conceivable route of administration, and a bewildering number of doses.

Several limitations on these models have become apparent with their use. First, in several respects, rodents show differential sensitivity to lead, as compared with man. For example, the rat appears more susceptible to cerebellar haemorrhage. Discrepancy in dose has provided some challenge to extrapolation from animals to man, and questions the realism of low level lead neurotoxicity, as demonstrated in animals (Michaelson, 1980). The relatively short life spans of experimental animals alter the relationship between dose and duration assumed in man. Moreover, comparing animals with man

Table 4 The administration of lead to rats and mice: reported models and doses

Reference	Dose	Source of lead	Route and duration of administration
Ahrens and Vistica, 1977	2, 3, 4%	Lead carbonate	In maternal diet from birth, *or*
	1.0, 1.3, 1.5%	Lead acetate	In maternal drink from birth
Alfano and Petit, 1981		As lead acetate	As Petit and Alfano, 1979
Alfano and Petit, 1982			As Petit and Alfano, 1979
Alfano et al., 1982			As Petit and Alfano, 1979
Angell and Weiss, 1982	0.2%	Lead acetate	In drink before or after weaning
Averill and Needleman, 1980	0.04, 0.4, 1, 2, 3%	Lead carbonate	In maternal diet from birth
Barlow et al., 1977	2%	Lead acetate	In maternal diet from birth
Barrett and Livesey, 1982	1 g/l	Lead acetate	In maternal drink from day 1 till day 18
Baumeister, 1982	100 mg/l	Lead acetate	In maternal drink from conception till birth
Brady et al., 1975	500 mg/kg	Lead acetate	p.o. to parent rats 30–90 days of age *and*
			To maternal rats till weaning
Brashear et al., 1978	1 g/kg	As lead acetate	p.o. to neonate rats on days 2–3, then bidaily thereafter
Brown, 1975	17.5, 25, 35 mg/kg	Lead acetate	p.o. to maternal rats from birth till day 20
Bull et al., 1979	5, 30, 200 mg/l	As lead chloride	In maternal drink from 14 days pregestation till weaning
Burright et al., 1982			
Campbell et al., 1982	0.2%	Lead acetate	As Dolinsky et al., 1981
Carmichael et al., 1981	300, 600, 1000, 2000, 3000, 4000 ppm	As lead acetate	In maternal drink from birth
Carmichael et al., 1982	300, 1000 ppm	As lead acetate	In maternal drink from conception to weaning
Carroll et al., 1977		As lead acetate	As Carmichael et al., 1981
Caruso et al., 1981			As Silbergeld and Goldberg, 1973
Clasen et al., 1974	2, 4%	Lead subacetate	As Zenick et al., 1981
			In maternal diet from birth
Cory-Slechta and Thompson, 1979	50, 300, 1000 ppm	As lead acetate	In drink to neonates from weaning till day 62
Cory-Slechta et al., 1981	100, 300 ppm	As lead acetate	In drink to neonates from weaning
Cramer et al., 1980	2.5 mg/ml	Lead acetate	In drink from birth
Crofton et al., 1980			As McCauley et al., 1979
Cutler, 1977	0.1%	Lead acetate	In drink from birth
Dietz et al., 1978	200 mg/kg	Lead acetate	p.o. to neonates from day 3 to day 30
Dolinsky et al., 1981	0.5%	Lead acetate	In drink from mating (in mice)
Drew et al., 1979	0.1%	Lead acetate	In maternal diet from birth to day 16, *then*
	10 ppm	As lead acetate	In diet from day 16 till day 21

Table 4 *continued*

Reference	Dose	Source of lead	Route and duration of administration
Driscoll and Stegner, 1976	0.01, 0.0001 mol/l	Lead acetate	In drink from mating till end of experiment
Driscoll and Stegner, 1978		As lead acetate	As Driscoll and Stegner, 1976
Dubas and Hrdina, 1978	50 µg	As lead acetate	p.o. to neonate rats from day 1–21, *then*
	80 ppm	As lead acetate	In drink
Dubas et al., 1978	2%	As lead acetate	In maternal drink from birth till day 21, *then*
	20, 40, 80 ppm	As lead acetate	In drink
Feeley et al., 1979		Lead acetate	As Zenick et al., 1979b
Flynn, 1979		Lead acetate	As Flynn et al., 1979
Flynn et al., 1979	0.5%	Lead acetate	In maternal drink, *or*
	0.2%	Lead acetate	In maternal drink, *and*
	225 µg/kg	Lead acetate	p.o. to neonate rats from birth till weaning, *then*
	0.25%	Lead acetate	In drink to neonates from weaning
Fowler et al., 1980			As Grant et al., 1980
Fox et al., 1977	0.2%	Lead acetate	In maternal drink from birth till weaning
Fox et al., 1979a			As Fox et al., 1977
Fox et al., 1979b	0.02, 0.2%	Lead acetate	In maternal drink from birth till weaning
Gatzke, 1980			As Winneke et al., 1977
Geiste and Mattes, 1979	25, 50 ppm	Lead acetate	In drink to neonates from day 23
Geist and Balko, 1980			As Geist and Mattes, 1979
Geist and Praed, 1982	19, 38 ppm	As lead acetate	In drink from conception
Gelman and Michaelson, 1979	25, 75, 225 mg/kg	As lead acetate	p.o. to neonates from day 2 till day 20
Gelman et al., 1979	10, 50, 225 mg/kg	As lead acetate	p.o. to neonates from day 2 till day 20
Gerber et al., 1978	1%	Lead acetate	In diet from 0.5 till 12 months, *or*
	0.1, 1 mg/ml	Lead acetate	In drink from birth till 1 month (in mice)
Goldman et al., 1982	0.05, 0.1, 0.2%	Lead acetate	In maternal drink from birth till weaning
Goldstein et al., 1974	4%	Lead carbonate	In maternal diet from day 12
Goldstein and Diamond, 1974			As Goldstein et al., 1974
Golter and Michaelson, 1975	1.09 mg	As lead acetate	p.o. to neonates from birth till day 16, *then*
	40 ppm	As lead acetate	In diet, *or*
	5%	Lead acetate	In maternal diet from birth till day 16, *then*
	40 ppm	As lead	In diet

Table 4 *continued*

Reference	Dose	Source of lead	Route and duration of administration
Govoni et al., 1978	2.5 mg/ml	Lead acetate	In maternal drink from gestation day 16
Govoni et al., 1979			As Govoni et al., 1978
Govoni et al., 1980	0.04, 2.5 g/l	Lead acetate	In drink from gestation day 16
Grant et al., 1980			As Kimmel et al., 1980
Gross-Selbeck and Gross-Selbeck, 1980		Lead	To give maternal blood leads of 20 and 60 µg/100 ml
Gross-Selbeck and Gross-Selbeck, 1981	0.1%	Lead acetate	In diet from weaning, gestation and lactation till weaning
Harry et al., 1982	100 mg/kg	As lead acetate	p.o. to neonate rats from day 2 till 24
Hastings et al., 1976	0.02, 0.1%	Lead acetate	In maternal drink from birth till day 21
Hastings et al., 1977			As Hastings et al., 1976
Hastings et al., 1979			As Fox et al., 1979
Hejtmancik et al., 1982	0.2%	Lead acetate	In drink from birth till weaning
Holtzman and Hsu, 1976	4%	Lead carbonate	In diet from day 14
Holtzman et al., 1980			As Holtzman and Hsu, 1976
Holtzman et al., 1981	1%	Lead carbonate	In maternal diet from birth till 2 weeks, *then*
			As Holtzman and Hsu, 1976
Hsu, 1981	0.5%	Lead acetate	In diet from birth
Jason and Kellogg, 1977	25, 75 mg/kg	Lead acetate	p.o. to neonates from day 2 till day 14
Jason and Kellogg, 1981			As Jason and Kellogg, 1977
Jones, 1979	225 mg/kg	Lead acetate	p.o. to neonates from day 1 till day 20
Kihara et al., 1981	50, 250 mg/kg	Lead acetate	p.o. to maternal rats from gestation day 0 till day 28
Kimmel et al., 1980			As Grant et al., 1980
Kiraly and Jones, 1982	10 mg/ml	Lead acetate	In drink from day 2
Klein and Koch, 1981			As Brown, 1975
Kostas et al., 1976	0.05, 0.5, 5%	Lead acetate	In maternal diet from birth till day 21, *then*
	0.25, 2.5, 25 ppm	As lead acetate	In diet from day 21 till day 35
Kostas et al., 1978	0.1%	Lead acetate	In maternal diet from birth till day 18, *then*
	10 ppm	Lead acetate	In diet
Krall et al., 1972			As Pentschew and Garro, 1972
Krass et al., 1980	2260 ppm	As lead acetate	In diet from 60 days pregestation till 4 months from birth
	1.25 g/kg	Lead acetate	p.o. to maternal rats on day 9, *or*
Krehbiel et al., 1976	0.83 g/kg	Lead acetate	p.o. to maternal rats from day 2 till weaning

Table 4 *continued*

Reference	Dose	Source of lead	Route and duration of administration
Krigman et al., 1972			As Pentschew and Garro, 1966
Krigman et al., 1974a			As Pentschew and Garro, 1966
Krigman et al., 1974b			As Pentschew and Garro, 1966
Krigman and Hogan, 1974			As Pentschew and Garro, 1966
Krigman et al., 1980	200 mg/kg	Lead	p.o. to neonates from days 3–30, *or*
	50 ppm	As lead	In diet throughout life (750 days)
Lampert et al., 1967			As Pentschew and Garro, 1966
LeFauconnier et al., 1980	10–80 mg/kg	Lead acetate	i.p. to neonate rats daily from day 5, *or*
			As Pentschew and Garro, 1966
Louis-Ferdinand et al., 1978	7.5 mg/kg	Lead acetate	i.p. to neonate rats daily from birth to day 10
Lucchi et al., 1981			As Govoni et al., 1978
McCauley et al., 1979	200 ppm	As lead chloride	In maternal drink from 14 days pregestation till weaning
McCauley et al., 1982	30,200 ppm	As lead chloride	As Bull et al., 1979
McConnell and Berry, 1979	4%	Lead acetate	In maternal diet from birth
McIntosh et al., 1982	5 mg/ml	Lead acetate	In drink from weaning
Memo et al., 1980a			As Govoni et al., 1978
Memo et al., 1980b	0.04, 2.5 g/l	Lead acetate	In drink from gestation day 16 till 3 months
Memo et al., 1981			As Govoni et al., 1980
Michaelson, 1973	4.5%	Lead acetate	In diet from birth
Michaelson et al., 1974	1.09 mg	As lead acetate	p.o. to neonates from birth till day 20, *then*
	400 ppm	As lead	In diet
Michaelson and Sauerhoff, 1974a	5%	Lead acetate	In maternal diet from birth till day 16, *then*
	25 ppm	As lead acetate	In diet till the end of the experiment
Michaelson and Sauerhoff, 1974b			As Michaelson and Sauerhoff, 1974a
Michaelson and Bradbury, 1982	0.1%	Lead acetate	In drink from birth
Milar et al., 1981	25, 100, 200 mg/kg	As lead acetate	p.o. to neonates from day 3 till day 30
Miller et al., 1982	50, 75, 100 mg/kg	Lead acetate	p.o. to maternal rats from 21 days pregestation till birth
Minsker et al., 1979	5, 25 mg/kg	As lead nitrate	i.v. to maternal rats on gestation day 17, *or*
	5, 25 mg/kg	As lead nitrate	p.o. to maternal rats from birth till weaning
Minsker et al., 1982			As Minsker et al., 1979
Modak et al., 1975	1%	Lead acetate	In drink from birth

41

Table 4 *continued*

Reference	Dose	Source of lead	Route and duration of administration
Modak et al., 1978	0.25, 0.5, 1%	Lead acetate	In drink from birth
Mullenix, 1977	5 mg/ml	Lead acetate	In maternal drink from birth till day 10
Mullenix, 1980	1%	Lead carbonate	In maternal diet from birth till day 20, or
			As Mullenix, 1977
Nakagawa et al., 1980	25 mg/kg	Lead acetate	s.c. to neonates from weaning for 4 days
Nathanson, 1979	15 μmol	Lead nitrate	p.o. to neonates from day 2 till day 25, then
		Lead nitrate	In drink at a calculated dose similar to p.o. exposure
Overmann, 1977	10, 30, 90 mg/kg	Lead acetate	p.o. to neonate rats daily on days 3–21
Overmann et al., 1981	0.02, 0.2%	Lead acetate	In drink from birth till weaning
Patel et al., 1974a			As Michaelson, 1973
Patel et al., 1974b			As Michaelson, 1973
Pentschew and Garro, 1966	4%	Lead carbonate	In maternal diet from birth
Petit and Alfano, 1979	0.4, 4%	Lead carbonate	In maternal diet from birth
Petit and LeBoutillier, 1979			As Pentschew and Garro, 1966
Petrusz et al., 1979	25, 100, 200 mg/kg	Lead acetate	p.o. from birth till day 20
Piepho and Adams, 1976	27, 54 mg/kg	Lead acetate	p.o. neonates from day 2 till day 20
	600 mg/kg	Lead acetate	p.o. to neonate rats daily from birth till day 10
Press, 1977a			As Press, 1977a
Press, 1977b			As Press, 1977a
Press, 1977c			
Rafales et al., 1981	0.02%	Lead acetate	In maternal drink from birth till day 21, or
			In drink from birth
Ramsay et al., 1980	100 μg/g	Lead acetate	p.o. to neonates from day 2 till day 30
Reiter et al., 1975	5, 50 ppm	Lead acetate	In drink from 40 days pregestation till end of experiment
Reiter and Ash, 1976	5%	Lead carbonate	In maternal diet from birth till weaning, then
	50 ppm	Lead acetate	In drink till the end of the experiment
Reiter, 1977			As Reiter and Ash, 1976
Reyners et al., 1976	1.8%	Lead acetate	In maternal diet from gestation day 14 till 3 months
Reyners et al., 1978	0.5%	Lead acetate	In maternal diet from birth
Reyners et al., 1979	100, 1000, 5000, 10,000 ppm	Lead	In maternal diet from birth
Santos-Anderson et al., 1980	1000 ppm	Lead acetate	In maternal drink from birth till weaning

Table 4 *continued*

Reference	Dose	Source of lead	Route and duration of administration
Sauerhoff and Michaelson, 1973	4%	Lead carbonate	In maternal diet from birth till day 17, *then*
	40 ppm	Lead carbonate	In diet till end of experiment
Schumann, 1977	5%	Lead acetate	In maternal diet from birth, *or*
	5%	Lead acetate	In maternal diet from birth till day 18, *then*
	400 ppm	Lead	In diet, *or*
Shigeta et al., 1980	2, 10 mg/ml	Lead acetate	As Silbergeld and Goldberg, 1973
Shih and Hanin, 1977	2, 5 mg/ml	Lead acetate	In drink from birth till day 60
Shih and Hanin, 1978a			As Sauerhoff and Michaelson, 1973
Silbergeld and Chisholm, 1976	5 mg/ml	Lead acetate	As Sauerhoff and Michaelson, 1973
			As Silbergeld and Goldberg, 1973
Silbergeld and Goldberg, 1973	2, 5, 10 mg/ml	Lead acetate	In drink from birth
Silbergeld and Goldberg, 1974a			As Silbergeld and Goldberg, 1973
Silbergeld and Goldberg, 1974b			As Silbergeld and Goldberg, 1973
Silbergeld and Goldberg, 1975	5 mg/ml	Lead acetate	As Silbergeld and Goldberg, 1973
Silbergeld et al., 1979	5, 10 mg/ml	Lead acetate	In drink from day 2
Silbergeld et al., 1980	5 mg/ml	Lead acetate	In drink from birth
Snowdon, 1973	0.5, 0.8, 1.2 mg/100 g	Lead acetate	i.p. for 37 days to adults or neonates from day 22
Sobotka and Cook, 1974	9, 27, 81 mg	Lead acetate	p.o. to neonates from day 3 till day 21
Sobotka et al., 1975			As Sobotka and Cook, 1974
Srivastava and Thakur, 1981	250, 1000, 3000, 6000, 31,000 ppm	As lead	In maternal drink through gestation and lactation
Stephens and Gerber, 1981	0.01, 0.1%	Lead	In diet from conception
Taylor et al., 1982	400, 200 mg/l	Lead acetate	As Bull et al., 1979
Tennekoon et al., 1979	5 mg/ml		As Silbergeld and Goldberg, 1973
Tesh and Pritchard, 1980	15 to 50 mg/kg	Lead nitrate	i.v. to maternal rats on gestation day 17
Thomas et al., 1971	4.5%	Lead carbonate	In maternal diet from birth, *and*
	1%	Lead acetate	In maternal drink from birth
Thomas and Thomas, 1974			As Thomas et al., 1971
Toews, et al., 1978	1 mg/kg	As lead acetate	p.o. to neonate rats on days 5–6, *or*
	1 mg/kg	As lead acetate	p.o. as above, then bidaily till day 30
Toews, et al., 1980	100, 400 mg/kg	As lead acetate	p.o. to neonate rats from days 2–30

Table 4 *continued*

Reference	Dose	Source of lead	Route and duration of administration
Verlangieri, 1979	0.1 mg/ml	Lead chloride	In drink from 14 days pregestation till end of experiment
Vistica and Ahrens, 1977	1.0, 1.3, 1.5%	Lead acetate	In maternal drink from birth
Weinreich et al., 1977	0.5, 1.0, 1.5%	Lead acetate	In diet from birth
Wince et al., 1976			As Pentschew and Garro, 1966
Wince and Azzaro, 1977			As Sauerhoff and Michaelson, 1973
Wince and Azzaro, 1978			As Sauerhoff and Michaelson, 1973
Wince et al., 1980			As Sauerhoff and Michaelson, 1973
Winder et al., 1984	300, 1000 ppm	As lead acetate	In maternal drink from conception to weaning
Winder et al., 1984			As Winder et al., 1984
Winder et al., 1984			As Winder et al., 1984
Winder et al., 1984			As Winder et al., 1984
Winneke et al., 1977	0.138%	Lead acetate	In diet from 60 days before breeding
Winneke et al., 1982	80, 250, 750 ppm	Lead acetate	In diet from 50 days before breeding
Zenick et al., 1978	1000 mg/kg	Lead acetate	In drink on restricted daily water intake
Zenick et al., 1979a	200, 400 mg/kg	Lead acetate	In drink on restricted daily water intake
Zenick et al., 1979b	750 mg/kg	Lead acetate	In drink on restricted daily water intake
Zenick et al., 1981	0.1, 0.2%	Lead acetate	In maternal drink from birth till weaning
Zenick and Goldsmith, 1981	0.02, 0.5%	Lead acetate	In maternal drink from gestation day 1 till weaning, *or* In drink from gestation day 1
Zenick et al., 1982	0.1, 0.2%	Lead acetate	In drink from birth, *or*
	0.1%	Lead acetate	In maternal drink from birth till weaning
Zimmering et al., 1982	0.5%	Lead acetate	In drink

Abbreviations: p.o., per os (via gastric lavage); i.p., intraperitoneal; i.v., intravenous; s.c., subcutaneous injection. Unless mentioned otherwise, day xx indicates xx days after parturition.

solely on the basis of external dose may be misleading since there are significant species differences in absorption and retention of lead.

With high and intermediate dose levels, effects of lead on the brain may also be obscured by undernutrition, itself a powerful cause of morphological change in the developing brain (Balazs et al., 1979). Published studies using high lead loads generally report a reduction in developmental weight gain, and this phenomenon appears to occur with a wide range of doses and experimental conditions. Using pair-fed controls, Michaelson and Sauerhoff (1974b) showed this loss in weight gain to be a consequence of reduced food intake by the maternal rat – the lead-adulterated diet was unpalatable. Bornschein et al. (1977) estimated that 0.2% lead acetate (1090 ppm Pb) in the drinking water was the maximum dose at which nutritional effects could be ruled out. This was shown experimentally by Carmichael et al. (1981), where doses above 1000 ppm Pb administered in the drinking water of maternal rats from conception till weaning produced significant weight loss in maternal rats and growth rate retardation in neonates. At doses of 1000 ppm Pb and below, the effects of undernutrition were shown to be minimal. However, by increasing the sample size of the numbers of animals used, this minimal effect becomes significant, mainly due to a decrease in group variance. This indicates that the use of a dose level of 1000 ppm Pb is probably too high as a limit for non-specific effects (Winder et al., in preparation, 1984d).

Operationally, many experimental studies are now assumed to be of low level toxicity on the grounds that no signs of overt poisoning, such as significant retardation in growth rate, were observed at the doses used. It is conventionally conceded that the brain is protected against 'slight' nutritional deficits but whether this is the case with lead is not known. The rationale for this seems logical, but may be dubious on the grounds that undernutrition is itself produced from altered biological phenomena, and a cellular 'undernutrition' may manifest itself long before any weight decrease.

Models and blood lead levels

Early attempts to produce experimental models of lead toxicity used doses of lead so high as to make their relevance to the problem of human lead poisoning questionable. Nutritional deficits and somatic changes were induced, causing problems of interpretation. Overall, it was found very difficult to compare the findings of animal studies with those in man.

There is some superficial attraction in using administered dose as the basis of comparison between studies. Unfortunately this is impractical given the variation in protocols. As lead is cumulative, it is clear that duration of exposure is at least as important as dose. Other factors which might make major differences in toxic responses to nominally equivalent doses are: (1) route of exposure, (2) age of experimental animals, (3) anion used, (4)

45

nutritional composition of normal diet and (5) mineral content of diet and drinking water. This is further complicated when the dose is indirect i.e. *in utero* or via milk, when a range of maternal factors come into play.

The intestinal absorption of lead is also extremely variable, and can range from as much as 50% in the neonate (Momcilovic and Kostial, 1974) to as little as 0.1% in the adult (Kostial *et al.*, 1971). In comparison, human absorption has been estimated at 5–10% in the adult (Kehoe, 1960); and 20–50% in children (Alexander, 1974). Estimation of lead intake as μg Pb/kg/day in experimental animals is thus a highly inaccurate measure of absorbed dose on a comparative basis.

Blood lead measurement, with all its obvious limitations, thus remains the most reliable means of assessing lead absorption. As an index of exposure, it obviates the necessity of crude estimations of dose, by focusing attention on the internal lead load. Cross study correlations become tenable, and comparison of lead levels in animals with those in human infants becomes much simpler. In the study of 'subclinical lead neurotoxicity' it becomes essential to use a model which gives a blood lead in experimental animals of around 80 μg Pb/100 ml. This is considered the threshold above which 'unequivocal cases of lead poisoning' may occur in children (USCDC, 1978), and this type of model would therefore be relevant to human conditions. However, a study of the literature shows that this criterion has rarely been met (Table 5). Carmichael *et al.* (1981) studied the dose relationships between dose and blood lead concentration, and doses of 300 ppm and 1000 ppm Pb produced blood lead concentrations of 40 μg and 85 μg Pb/100 ml respectively. These corresponded roughly to those limits used in clinical recommendations (see e.g. USCDC, 1978: CEC, 1980). Other workers have been using doses of lead that give blood leads in this range (e.g. Gross-Selbeck and Gross-Selbeck, 1980; 1981; Winneke *et al.*, 1982), although the production and maintenance of steady state blood levels probably does not represent true subclinical intoxication. This is generally of an acute nature.

It is only relatively recently that blood lead has been consistently reported, thus facilitating cross study comparisons (see Table 5). The majority of studies on the morphological effects of lead in the developing CNS have used concentrations far in excess of 80 μg Pb/100 ml, producing blood lead levels wildly outside those encountered in paediatric clinical practice. It is obvious from Table 5 that the Pentschew and Garro (1966) regime produces blood leads in neonates too high to be of any relevance to the human conditions they are supposed to model (650 μg Pb/100 ml at 21 days, rising to 1297 μg Pb/100 ml at 30 days). Furthermore, although the use of pair-fed animals appears to control for the effects of undernutrition, it does not consider the interrelationships between lead and undernutrition, and therefore does not account for their combined effects (i.e. a double jeopardy). Only in studies reporting no weight differences between untreated and dosed animals can the possibility of concluding specific lead effects be entertained.

Table 5 The effect of various regimes of lead administration and age on blood lead

Reference	Dose	Anion	Route of administration	Period of dosing	Dietary concentration (ppm Pb)	Daily dose Pb (mg/kg)	Age (days) when blood sampled	Blood lead (μg/100 ml)
Winneke et al., 1982	80 ppm	Acetate	in food	from	80		steady	5
	250 ppm	Acetate	in food	50 days	250		state	11
	750 ppm	Acetate	in food	pregestation	750		levels	18
Bull et al., 1979	5 ppm	Chloride	in drink	from 14 days	5		21	12
	30 ppm	Chloride	in drink	progestation –	30		21	21
	200 ppm	Chloride	in drink	weaning	200		21	36
McCauley et al., 1982	30 ppm	Chloride	in drink	from 14 days	30		10	12
	30 ppm	Chloride	in drink	before breeding	30		21	17
	200 ppm	Chloride	in drink	till	200		10	40
	200 ppm	Chloride	in drink	weaning	200		21	42
Goldman et al., 1980	0.05%	Acetate	in drink	from birth	273		21	13
	0.1%	Acetate	in drink	from birth	545		21	21
	0.2%	Acetate	in drink	from birth	1090	0.5*	21	47
Grant et al., 1980	5 ppm	Acetate	in food	from	5		30	16
	25 ppm	Acetate	in food	45 days	25		30	18
	50 ppm	Acetate	in food	pregestation	50		30	48
Rafales et al., 1981	0.02%	Acetate	in drink	from birth	109		21	19
Hastings et al., 1979	0.02%	Acetate	in drink	from birth	109		20	24
	0.2%	Acetate	in drink	to weaning	1090	0.5*	20	69
Winneke et al., 1977	0.138%	Acetate	in food	60 days pregestation	745		16	27
Hastings et al., 1977	0.02%	Acetate	in drink	from birth	109		20	29
	0.1%	Acetate	in drink	to weaning	545		20	69
Krass et al., 1980	2260 ppm	Acetate	in food	60 days pregestation	2260		16	34
Hejtmancik et al., 1982	0.2%	Acetate	in drink	from birth	1090	0.5*	10	38
	0.2%	Acetate	in drink	till weaning	1090	0.5*	21	47
Carmichael et al., 1981	300 ppm	Acetate	in drink	from	300		21	40
	600 ppm	Acetate	in drink	conception	600		21	55
	1000 ppm	Acetate	in drink	to	1000	0.5*	21	85
	4000 ppm	Acetate	in drink	weaning	4000		21	194

Table 5 *continued*

Reference	Dose	Anion	Route of administration	Period of dosing	Dietary concentration (ppm Pb)	Daily dose Pb (mg/kg)	Age (days) when blood sampled	Blood lead (µg/100 ml)
Zenick et al., 1982	0.1%	Acetate	in drink	from birth	545		20	36
	0.2%	Acetate	in drink	from birth	1090	0.5*	20	54
Michaelson and Bradbury, 1982	0.1%	Acetate	in drink	from birth	545		21	37
Winder et al., 1984b	300 ppm	Acetate	in drink	from	300		10	47
	300 ppm	Acetate	in drink	conception	300		21	80
	1000 ppm	Acetate	in drink	to	1000	0.5*	10	72
	1000 ppm	Acetate	in drink	weaning	1000	0.5*	21	105
Fox et al., 1977	0.2%	Acetate	in drink	from birth	1090	0.5*	21	65
Mykkanen et al., 1982	0.5%	Acetate	in food	from birth	2635		90	76
	0.5%	Acetate	in food	till	2635		180	78
	0.5%	Acetate	in food	weaning	2635		360	103
Weinreich et al., 1977	0.5%	Acetate	in food	from birth	2725		52	145
	1.0%	Acetate	in food	from birth	5450		52	186
	1.5%	Acetate	in food	from birth	6175		52	163
Petrusz et al., 1979	25 mg/kg	Acetate	by gavage	from		25	15	203
	100 mg/kg	Acetate	by gavage	birth		100	15	1014
	200 mg/kg	Acetate	by gavage	to day 20		200	15	1056
McConnell and Berry, 1978	4%	Acetate	in food	from birth	21,840		25	258
Petit and Alfano, 1979	0.4%	Carbonate	in food	from birth	3204		30	331
Schumann, 1977	4%	Carbonate	in food	from birth	32,040		30	1297
	5%	Acetate	in diet	from birth	27,300		peak	354
Averill and Needleman, 1980	2%	Carbonate	in food	from birth	8010		21	385
LeFaucomier et al., 1980	10 µg/kg	Acetate	i.p. injection	daily		10	21	650
	40 µg/kg	Acetate	i.p. injection	from day 5		40	9	570
Barlow et al., 1977	4%	Carbonate	in food	from birth	32,040	1.5*	21	650
	2%	Acetate	in food	from birth	10,920		1–5	44
Press, 1977	2%	Acetate	in food	from birth	10,920		20–22	1240
	328 µg/kg	Acetate	by gavage	from birth		328	2	608
	328 µg/kg	Acetate	by gavage	till day 10		328	10	1839

*Estimations of daily lead dose from Bornschein, Fox and Michaelson (1977)

48

The distribution of lead in the brain

Early studies showed that lead was present in the brain of lead-dosed animals in appreciable quantities (e.g. Sauerhoff and Michaelson, 1973). The accumulation of lead in the brain is related to blood lead levels (Goldstein *et al.*, 1974), although the levels of lead within different brain areas show considerable variability.

Generally, forebrain or cerebral cortex shows increased lead concentration compared with other brain areas. The exception to this is in the presence of cerebellar haemorrhage, when cerebellar lead concentrations increase considerably. The raised level of lead in haemorrhagic regions of the CNS is due however to the presence of erythrocytes in damaged areas. By using a double labelling technique in heavily lead-dosed animals with petechial brain haemorrhage, Kochen and Greener (1977) showed that 63% of cerebellar lead was present in blood cells, compared with 29% in forebrain. By correcting for blood lead concentrations the figures for cerebellar and forebrain tissue lead became similar.

Studies on the regional distribution of lead in the CNS have given conflicting results. Initially, Fjerdingstad and colleagues studied aspects of lead uptake in brain, and reported that lead accumulated in the hippocampus of normal rats to a concentration seven times that of the entire brain (Fjerdingstad *et al.*, 1974b). Hippocampal lead was said to account for about half the lead in the brain. Danscher *et al.* (1975) studied the accumulation of lead in the amygdala, and found a comparably high concentration (although slightly less than that found in the hippocampus) accounting for about 14% of the total lead content of the brain. Danscher *et al.* (1976) studied the dry weight distribution of lead in the hippocampus, subdividing it into four regions (CA1, CA3, hilus of dentate fascia and molecular layer of dentate fascia). There was a gradient in lead content from CA1 (25 ppm Pb), CA3 (30 ppm Pb), molecular layer (40 ppm Pb) and hilus (60 ppm Pb). As a whole, the hilus contained three times the amounts of lead found in the hippocampus. It should be noted that these three studies were carried out in normal non-dosed animals, although the dietary lead content seems relatively high (0.235–0.760 ppm Pb). A later multielemental analysis of rat hippocampus using proton-induced X-ray emission spectroscopy reported that the concentration of lead in the hippocampus may be much lower, giving a mean of 3 ppm Pb in non-fixed sections from normal rats (Kemp and Danscher, 1979). This is still a controversial area of research.

Subsequent studies in lead-dosed animals give conflicting evidence concerning the selective accumulation of lead in the hippocampus. Klein and Koch (1981) reported that rats dosed intraperitoneally with 5 and 7.5 mg/kg lead from days 1–10 had more lead in several brain regions than those dosed from days 11–20. The highest concentration was found in the cerebellum, and the lowest in the brain stem and hippocampus, with

Table 6 The uptake of lead into the CNS

Ref.	Route	Dose	Duration	Area of brain									Blood
				BS	PM	CB	MdB	Hth	Str	Hip	FB	Cort	
1	p.o.	50	1–21 days		4.21	4.59	n.d.	3.76	2.11	2.62		1.28	
	drink	80 ppm	8 weeks		3.51	2.12	1.27	2.88	2.65	3.48		1.77	
2	i.p.	5.0	1–10 days	0.145*		0.435						0.232	25.3
	i.p.	7.5	1–10 days	0.128*		0.512						0.316	35.3
	i.p.	5.0	11–20 days	0.167*		0.385						0.216	21.4
	i.p.	7.5	11–20 days	0.215*		0.390						0.293	31.0
3	diet	0.5%	3 months	0.90		0.80					1.48*		76
	diet	0.5%	6 months	0.70		0.70					1.45*		78
	diet	0.5%	12 months	1.05		1.14					2.02*		108
4	p.o.	0.005	8 weeks				0.020		0.035	0.029		0.036	
	p.o.	0.025	8 weeks				0.025		0.037	0.074		0.035	
	p.o.	0.1	4 weeks				0.022		0.038	0.101		0.038	
	p.o.	0.1	6 weeks				0.062		0.105	0.110		0.052	13.8
	p.o.	0.1	8 weeks				0.078		0.075	0.101		0.062	
	p.o.	1.0	8 weeks				0.175		0.180	0.200		0.175	31.1
5	drink	0.2%	8 months		0.50	0.56	0.50	0.48	0.56	1.00		1.42	

References: (1) Hrdina et al. (1980), (2) Klein and Koch (1981), (3) Mykkanen et al. (1982), (4) Collins et al. (1982) and (5) LeFauconnier et al. (1983)
Abbreviations: Route: i.p., intraperitoneally (mg/kg); p.o., by gavage (mg/kg). Area: BS, brain stem; PM, pons-medulla; CB, cerebellum; MdB, midbrain; Hth, hypothalamus; Str, striatum; Hip, hippocampus; Fb, forebrain; Cort, cerebral cortex; n.d., not detectable.
*The description of the dissection of these areas indicates presence of the hippocampus.
Units: All brain area lead levels are expressed as μg Pb/g wet weight. Blood lead concentration measured as μg Pb/100 ml

intermediate levels in cerebral cortex. Mykkanen *et al.* (1982) found the highest concentration of lead to be in the forebrain (including the hippo-campus). LeFauconnier *et al.* (1983) noted that lead concentration in the cerebral cortex (1.4 mg/g) was higher than in the hippocampus (1.0 mg/g), both values being nearly double those found elsewhere in the CNS. These two latter studies used longterm dietary exposure to lead (2625 ppm Pb in the drink for 12 months in the first, and 1090 ppm Pb also in the drink for 8 months in the second) and their findings do not exclude the possibility of short term and perhaps selective sequestration. This possibility is raised by the study of Collins *et al.* (1982), where intermediate length exposure by gavage was used. These authors reported that more lead appeared to be deposited in the hippocampus than any other brain region examined (cortex, striatum, midbrain) at 4 weeks and with relatively low dose. The concentrations of lead in various CNS regions reported in these studies are shown in Table 6.

The disposition of lead within cells has also received some attention, since intranuclear inclusion bodies were reported in the kidney by Blackman (1936). These have also been found in the nervous system. Shirabe and Hirano (1977) reported that various cytoplasmic and intranuclear inclusions were found in macrophages and astrocytes of rat brains in which pellets of lead acetate had been implanted. These inclusions were found to contain varying proportions of lead by X-ray microanalysis. Lead inclusion bodies have also been found in the anterior horn cells of the cervical spinal cord and neurones of the substantia nigra in monkeys chronically exposed to lead giving mean blood leads ranging up to 95 μg Pb/100 ml (Osherhoff *et al.*, 1982). Silberdeld and Adler (1977) and Silbergeld *et al.* (1977) noted that lead appeared in significant quantities in mitochondria, suggesting a possible locus for neurochemical action.

4
The Behavioural Effects of Lead

It has been suggested that ambient lead levels are sufficiently high in urban environments to cause impairment of brain function in children whose lead load is at the higher end of the normal range. Some authors believe that lead may be a contributory factor in childhood hyperactivity (David *et al.*, 1972). Others have claimed that children with relatively high, though not overtly toxic, blood lead levels can be distinguished from their low lead peers by measures of intelligence and classroom performance (Needleman *et al.*, 1979; Yule *et al.*, 1981). This topic has been the subject of several detailed and authoritative reviews, the best of these being by Rutter (1980) and Bornschein *et al.* (1980a).

EXPERIMENTAL APPROACHES

The investigation of the behavioural effects of lead in animals has drawn on all aspects of experimental psychology. Only studies employing rats or mice will be discussed here, although a large body of evidence is available for other animals (including fish, birds, sheep and primates). The behavioural effects of developmental lead exposure are summarized in Table 7. The doses and routes of administration for the studies outlined here are shown in Table 7.

As shown in Table 7 the behavioural experiments can be divided into four main types, (a) simple behaviour and activity, (b) learning paradigms, (c) complex behaviour and (d) drug-induced behaviour.

SIMPLE BEHAVIOUR AND ACTIVITY

Following the proposed links between lead and hyperactivity in children, attempts were made to correlate lead exposure to hyperactivity in experimental animals. The first study to report such a correlation was that of Silbergeld and Goldberg (1973; 1974a and b; 1975) in mice given lead at doses of up to 10 mg Pb/l in drinking water (5650 ppm Pb). The hyperactivity observed was reproducible, although above 5 mg Pb/l (2565 ppm Pb), dosed animals were underweight.

Table 7 Behavioural effects of developmental lead exposure in rats and mice

Table 7a: Simple behaviour and activity

Behavioural test and parameter	Result or effect	Reference
Eye opening	Delayed (in mice)	Silbergeld and Goldberg, 1973; 1974b
	Delayed	Reiter and Ash, 1976
Appearance of hair	Delayed (in mice)	Silbergeld and Goldberg, 1973; 1974b
Righting reflex	Delayed	Reiter and Ash, 1976
	Longer duration	Fox et al., 1979
Co-ordinated walking	Delayed (in mice)	Silbergeld and Goldberg, 1973; 1974b
Startle response	Delayed	Reiter and Ash, 1976
Developmental landmarks	No change	Dubas and Hrdina, 1978
Swimming ability	Not altered (between days 6–24)	Overman et al., 1981
	No change	Sobotka et al., 1975
General motor activity	Increased (in mice)	Silbergeld and Goldberg, 1973; 1974a; b
Activity	Increased	Michaelson et al., 1974
	Increased	Golter and Michaelson, 1975
	No change	Wince et al.,1976
	Increased	Schumann, 1977
	Increased	Wince et al., 1980
	Inferior	Lucchi et al., 1981
Locomotor co-ordination		Tesh and Pritchard, 1980
Home cage — emotionality	Consummatory behaviours increased (in mice)	Zimmering et al., 1982
Open field — activity	Increased	Geist and Balko, 1980
	Not altered	Geist and Praed, 1982
	Increased	Winneke et al., 1977
	Increased	Driscoll and Stegner, 1978
	Increased	Govoni et al., 1978
	Increased	Govoni et al., 1979
	More active	Govoni et al., 1980
	No change	Petit and Alfano, 1979
	Hypoactivity	Grant et al., 1980
		Krass et al., 1980

Table 7 *continued*

Behavioural test and parameter		Result or effect	Reference
		Increased during dark phase (in mice)	Dolinsky *et al.*, 1981
		No change during light phase (in mice)	Gross-Selbeck, 1981
		No change	Barrett and Livesey, 1982
		No change	Minsker *et al.*, 1979; 1982
		No change	Geist and Praed, 1982
	rearing	Increased	Winneke *et al.*, 1977
		Decreased	Kihara *et al.*, 1981
		No change	Geist and Praed, 1982
	grooming	Increased	Winneke *et al.*, 1977
		Reduced	Geist and Praed, 1982
	jumping	Increase (in mice)	Zimmering *et al.*, 1982
Running wheel	activity	No change	Kostas *et al.*, 1976
		No change	Hastings *et al.*, 1976
		Difference in initial stages	Hastings *et al.*, 1979
		Hypoactivity	Verlangieri, 1979
Residential maze	activity	Hypoactivity	Reiter *et al.*, 1975
		No change	Reiter and Ash, 1976
		Initial hyperactivity, disappearing by day 44	Reiter, 1977
Rotarod	activity	Impairment in co-ordination	Overmann, 1977
		No change	Cramer *et al.*, 1980
		No change	Grant *et al.*, 1980
		Decreased time at high dose	Kihara *et al.*, 1981
		No change	Minsker *et al.*, 1979; 1982
		Decreased	Overman *et al.*, 1981
Photoactivity cage	activity	No change	Krehbiel *et al.*, 1976
		Increased at high dose	Gelman *et al.*, 1979
Selective activity meter		Hyperactivity	Sauerhoff and Michaelson, 1973
		More aggressive	
		Excessive stereotyped self grooming	
Jiggle cage	activity	Increased at 2 weeks	Reiter and Ash, 1976

55

Table 7 *continued*

Behavioural test and parameter		Result or effect	Reference
Plexiglass activity cage (motor acts)			
	frequency and duration	No change at 6 weeks	Overmann, 1977
	patterning of behaviour	Increased	Reiter, 1977
Circular photoelectric cage		Not affected	Mullenix, 1977
Photocell chamber	activity	Affected	Mullenix, 1977
		Frequency of specific clusters altered	Mullenix, 1980
'Animex' electronic activity cage		No change	Cramer et al., 1980
		Increased	Rafales et al., 1981
		No change	Zenick et al., 1982
'Varimex' activity meter activity		Increased	Weinreich et al., 1977
		Increased	Memo et al., 1980a
		Increased	Lucchi et al., 1981
Infrared photodetector array	horizontal activity	Hyperactivity at 6–8 weeks, normal at 12 weeks	Dubas and Hrdina, 1978
Home cage exploration apparatus	locomotor activity	As Dubas and Hrdina, 1978	Dubas et al., 1978
	exploratory activity	Elevated	Nathanson et al., 1979
Enriched or isolated environments	activity	Delayed maturation	Crofton et al., 1980
	passive avoidance latency	Delayed maturation	Crofton et al., 1980
Activity chamber		Increased	Petit and Alfano, 1979
		Decreased	Petit and Alfano, 1979
		Hyperactivity at 18 days	Jason and Kellogg, 1981
Table 7b Learning paradigms			
Appetitive learning	acquisition	No change	Taylor et al., 1982
	extinction	Increased	Taylor et al., 1982
Discrimination tasks			
visual	acquisition	Slower	Driscoll and Stegner, 1976
	reversal	No change	Hastings et al., 1976
		No change	Driscoll and Stegner, 1976

56

Table 7 *continued*

Behavioural test and parameter		Result or effect	Reference
orientation (simple)		No change in performance	Winneke et al., 1977
size discrimination (difficult)		Very inferior performance	Winneke et al., 1977
Lashley jumping stand		Increase of error repetitions	Krass et al., 1980
	simultaneous	Inferior performance	Winneke et al., 1982
	successive	Delay in reaching criterion	Hastings et al., 1979
		No change	Hastings et al., 1977
		No change	Hastings et al., 1979
Tactually cued		Impaired performance	Overmann, 1977
	Go/No Go	No change	Hastings et al., 1979
Simple learning			
Learning ability			
Hebbs–Williams closed field maze		Inferior	Tesh and Pritchard, 1980
		No change	Snowdon, 1973
		Impaired learning ability	Geist and Mattes, 1979
		No change	Petit and Alfano, 1979
T-maze	acquisition	Decreased	Brown, 1975
Water T-maze	black/white	More errors, longer swimming times	Brady et al., 1975
	brightness	More errors, longer swimming times	Zenick et al., 1978
	shape	More errors, longer swimming times	Zenick et al., 1978
Underwater T-maze	swimming time	Greater at high dose	Kihara et al., 1981
E-maze	acquisition	No change	Overmann, 1977
	reversal	No change	Overmann, 1977
Radial maze	activity	No change	Flynn et al., 1979; Flynn, 1979
	spontaneous alternation	No change	Flynn et al., 1979; Flynn, 1979
Radial 8-arm maze	acquisition	Delay	Alfano and Petit, 1981
Spontaneous alternation		reduced alternation	Jones, 1979
Spatial alternation		Lower rate, higher total arm entries	Kostas et al., 1976
		No change	Milar et al., 1981
Shuttle apparatus	crossings	Increased	Kostas et al., 1976
Shuttle box		No change	Zimmering et al., 1982
one-way shuttle task	acquisition	No change	Sobotka et al., 1975
two way shuttle avoidance task		Deficit in performance	Sobotka and Cook, 1974

Table 7 *continued*

Behavioural test and parameter		Result or effect	Reference
	acquisition	Impaired	Sobotka et al., 1975
	performance	Improved performance	Driscoll and Stegner, 1976
		No change	Alfano and Petit, 1981
	reversal	No change	Alfano and Petit, 1981
		Improved performance	Winneke et al., 1982
Avoidance/escape conditioning task	intertrial crosses	Increased perseverance of habit responding	Sobotka et al., 1975
		Decreased	Winneke et al., 1982
Two-way active avoidance	acquisition	No change	Sobotka et al., 1975
	reversal	No change	Alfano and Petit, 1981
Passive shock avoidance	acquisition	No change	Alfano and Petit, 1981
	reversal	No change	Minsker et al., 1979; 1982
	acquisition	No change	Overmann, 1977
Single lever response		No change	Overmann, 1977
Operant delayed response		No change	Sobotka et al., 1975
Fixed interval schedule		Increased response rates	Overman, 1977
	interresponse time	Poorer performance	Cory-Slechta and Thompson, 1979
		Lengthened by postweaning exposure	Zenick et al., 1979b
		Shortened by preweaning exposure	Angell and Weiss, 1982
Fixed ratio schedule	latency	Intersubject variability	Cory-Slechta and Thompson, 1979
		Altered	Padich and Zenick, 1977
	interresponse time	Lengthened by postweaning exposure / Shortened by preweaning exposure	Angell and Weiss, 1982
Continuous reinforcement schedule		No change	Shigeta et al., 1980
Sidman avoidance schedule		No change	Zenick et al., 1979b
		Poorer performance	Krigman et al., 1972
		No change	Shigeta et al., 1980
Differential reinforcement low rate	interresponse time	Increased number	Dietz et al., 1978
	response latency	Increased	Cory-Slechta et al., 1981
	response duration	Shortened	Cory-Slechta et al., 1981

Table 7 *continued*

Behavioural test and parameter		Result or effect	Reference
high rate	performance	Deficient	Alfano and Petit, 1981
	performance	Increased at low dose	Gross-Selbeck and Gross-Selbeck, 1980; 1981
		Decreased at high dose	Gross-Selbeck and Gross-Selbeck, 1980
Table 7c Complex behaviour			
Squealing		Increased	Kostas *et al.*, 1976
Flinch–jump test		No change	Hastings *et al.*, 1977
Visual evoked response	primary P1N1 component	Increase in latency	Fox *et al.*, 1977
	secondary P2N2 component	Increase in latency	Fox *et al.*, 1977
	CNS 'recoverability' function	Decreased	Fox *et al.*, 1977
	P2 component	Altered latency	Feeley *et al.*, 1979
Shock-elicited aggression test		Decreased	Hastings *et al.*, 1976
		No change	Hastings *et al.*, 1977
Tonic seizure		Earlier ontogenetic appearance	Fox *et al.*, 1979
Electroshock seizure threshold		Unaffected	Overman *et al.*, 1981
Maximal electroshock seizure		Disturbed	Fox *et al.*, 1979
Audiogenic seizure		No change	Alfano and Petit, 1981
Seizures		Increased (in mice)	Burright *et al.*, 1982
Transcorneal ECS		Increased (in mice)	Burright *et al.*, 1982
Reproductive test		Decreased mating index	Kihara *et al.*, 1981
Maternal behaviour		No effect	Zenick *et al.*, 1979a
Social behaviour		Reduced social and sexual investigation (in mice)	Cutler, 1977
		reduced agonistic behaviour (in mice)	Cutler, 1977
Agonistic behaviour		Altered	Burright *et al.*, 1983
Table 7d Drug induced behaviour			
Amphetamine	spontaneous alternation	Paradoxically decreased	Kostas *et al.*, 1978
	discrimination learning	Less sensitive	Zenick and Goldsmith, 1981
	DRL	Increased variability	Dietz *et al.*, 1978
	two-way shuttle	Alleviated deficit	Sobotka and Cook, 1974

59

Table 7 continued

Behavioural test and parameter		Result or effect	Reference
	motor activity	Curtailed at high dose	Sobotka and Cook, 1974
	activity	Paradoxically decreased (in mice)	Silbergeld and Goldberg, 1973; 1974a; b
		Dose related diminution at 50 ppm Pb	Reiter et al., 1975
		Attenuated increased	Reiter and Ash, 1976
		Attenuated increased	Wince et al., 1976
		Increased	Kostas et al., 1978
		Attenuated	Wince et al., 1980
		Paradoxical decrease	Memo et al., 1980a
		Attenuated increase at high dose	Caruso et al., 1981
		No change during dark phase (in mice)	Dolinsky et al., 1981
		Decreased during light phase (in mice)	
		Decreased at low, increased at high dose	
		21% increase in males	Jason and Kellogg, 1981
		Attenuated increased	Rafales et al., 1981
		Attenuated response overall	Zenick et al., 1981
		females more active	Zenick et al., 1982
	grooming	Decreased	Kostas et al., 1978
	seizures	Attenuated (in mice)	Burright et al., 1982
Methylphenidate	activity	Paradoxically increased (in mice)	Silbergeld and Goldberg, 1974a; b; 1975
Apomorphine	aggression	Attenuated	Drew et al., 1979
	stereotypy	Unaffected	Drew et al., 1979
	activity	Increased plus circling (in mice)	Silbergeld and Goldberg, 1975
	activity	Attenuated	Wince et al., 1980
L-DOPA	activity	Increased (in mice)	Silbergeld and Goldberg, 1975
α-MPT	activity	Decreased (in mice)	Silbergeld and Goldberg, 1975
α-MPT + amphetamine	activity	Increased (in mice)	Silbergeld and Goldberg, 1975
Cocaine	seizures	Attenuated (in mice)	Burright et al., 1982
Fenfluramine	activity	Paradoxically decreased (in mice)	Silbergeld and Goldberg, 1975
(−)Sulpiride	activity	Decreased below control	Lucchi et al., 1981
Haloperidol	activity	Slight decrease	Lucchi et al., 1981
Physostigmine	activity	Decreased (in mice)	Silbergeld and Goldberg, 1975

Table 7 *continued*

Behavioural test and parameter		Result or effect	Reference
Neostigmine	activity	No change (in mice)	Silbergeld and Goldberg, 1975
Dimethylaminoethanol	activity	Paradoxically decreased (in mice)	Silbergeld and Goldberg, 1975
Atropine	activity	Increased responsiveness (in mice)	Silbergeld and Goldberg, 1975
		No change	Cramer *et al.*, 1980
Benztropine	activity	Increased (in mice)	Silbergeld and Goldberg, 1975
Pentobarbitol	activity	Paradoxically increased (in mice)	Silbergeld and Goldberg, 1974a; b
	DRL	No change	Dietz *et al.*, 1978
Chloral hydrate	sedation	No change (in mice)	Silbergeld and Goldberg, 1974a; b
Chlorpromazine	sedation	Decreased (in mice)	Silbergeld and Goldberg, 1975
Picrotoxin	convulsions	Occurred at lower dose of drug	Silbergeld *et al.*, 1979
Isoniazid	convulsions	Occurred at lower dose of drug	Silbergeld *et al.*, 1979
Penetylenetrazol	convulsions	No change	Silbergeld *et al.*, 1979
Strychnine	convulsions	Occurred at lower dose of drug	Silbergeld *et al.*, 1979
Mercaptopropionic acid	clonus	Occurred at lower dose of drug	Silbergeld *et al.*, 1979

Since then there have been many studies on activity measures in lead-dosed rats. These have used all manner of devices, from simple test systems such as the open field or running wheel, through to automated activity chambers (see Table 7a).

The open field has been used often in studying the behavioural effects of lead on activity, although this paradigm is more properly a measure of emotionality (see, e.g. Archer, 1973; Walsh and Cummins, 1976). Other measures include rearing, sniffing and grooming. These may or may not be affected by lead (see Table 7a). Geist and Balko (1980) reported that emotionality (as measured by urination and defaecation) was increased in rats dosed with 50 ppm lead acetate. No changes were seen in emotionality, locomotor activity or rearing at 19 ppm and 38 ppm Pb, although a decrease in grooming was seen at the higher dose (Geist and Praed, 1982). Zenick *et al.* (1979e) correlated open field activity to reduce birth and/or weaning weights. Lead appears to be a confounding, rather than a direct factor.

General effects of lead on neurobehavioural development were investigated in a two generation rat study (Reiter *et al.*, 1975; Cahill *et al.*, 1976). Rats were given 5 or 50 ppm lead as the acetate in drinking water for 40 days prior to mating, after which treatment of pregnant females was continued throughout gestation and lactation, and the offspring (F1) and their progeny (F2) were similarly exposed from weaning throughout adulthood (F1 maternal rats and F2 female offspring were removed from the study at weaning). Blood lead levels in the F1 maternal rats were 11, 13 and 26 μg/100 ml respectively in the control, 5 and 50 ppm groups, while the F2 neonates showed levels of 8, 11 and 20 μg/100 ml at birth and 5, 6 and 10 μg/100 ml at 180 days. Mean body weights of the F2 offspring during the development period were not significantly lower than controls, and indeed males at 50 ppm were significantly heavier on days 15 and 18. However, relative brain weight of the offspring at birth was significantly reduced at both levels. Significant delays were noted in the development of the righting reflex at 5 and 50 ppm, but there was no difference in the development of the startle response. Adult males at both levels showed a significant reduction in locomotor activity in a residential maze, and did not respond to amphetamine injection with the normal degree of hyperactivity. EEG measurements on anaesthetized F1 adult males showed changes at the 5 ppm level that were consistent with reduced activity. However, no such effect was seen in the 50 ppm F1 adults, nor in the F2 adults at either level. Other studies in rats have also suggested that low level lead treatment reduces rather than increases activity (Sobotka and Cook, 1974; Sobotka *et al.*, 1975) in contrast with the suggestion noted above that in children excessive lead exposure may be associated with hyperactivity.

Behavioural changes were also evident in rats that had been exposed to lead only via maternal milk. Maternal rats were given drinking water containing 0.02% lead acetate (109 ppm Pb) from parturition to weaning at 21 days, after which the neonates were given a normal diet and tap water. At 20 days the milk contained a mean level of 21 μg Pb/100 ml, whereas mean blood lead concentrations in the offspring were 29, 5 and 9 μg/100 ml at 20, 60 and 270 days respectively (11, 4 and 9 μg/100 ml in control rats). Brain lead levels were similar to blood lead levels at 29 days, but in both groups were higher than blood lead levels at 60 and 270 days. When the offspring were tested for shock-elicited aggression at 60 days of age, the lead-exposed group showed significantly less aggressive behaviour. However, they did not differ in their acquisition and subsequent reversal of a successive brightness discrimination task (see below), on which they were started at approximately 90 days of age. Wheel running activity was not assessed in these rats, but another group of offspring from maternal rats given the higher level of 1000 ppm lead acetate showed less activity than controls only in the first 2 h. Their activity over the remainder of the 21 day test period (beginning at 30 days of age) was unaffected (Hastings et al., 1977).

In a comprehensive study involving the administration of 0.5, 5, 25, 50 or 250 ppm lead as the acetate in drinking water to mothers for 6–7 weeks before mating, and to their offspring for up to 9 months after birth, the 25 ppm level was sufficient to affect performance of an operant task in adults (see below), and the development of surface and air righting in the offspring was delayed at 50 and 250 ppm. Locomotor development was unaffected except for an increase in pivoting at 14 days of age in animals given 250 ppm, and post-weaning activity levels were altered. Motor co-ordination in a rotorod test was also unchanged. However, these rats were fed a semi-purified diet and not a normal diet as in most similar studies. Gastro-intestinal absorption of the lead in the drinking water would be expected to be facilitated by such a regime, and this is evidenced by the high blood lead levels attained. Median blood lead levels in the offspring given 0, 25 and 50 ppm lead were 4, 37 and 57 μg/100 ml at day 1 and had fallen to 3, 22 and 35 μg/100 ml respectively by day 11 (Grant et al., 1980). The blood levels attained were sufficient to affect other systems beside the CNS; the age of vaginal opening was significantly delayed at 25 ppm or more, increased amino laevulinic acid (ALA) excretion and kidney damage were evident even at 5 ppm, and kidney weights were increased at 0.5 ppm or more (Fowler et al., 1980).

A characteristic behaviour shown by rats is a class of exploratory activity known as spontaneous alternation. This is normally measured in a symmetrical T-, Y- or E-maze, and the animal is allowed to freely explore the apparatus on successive trials. The rat tends to distribute its responses in a non-random manner, visiting the most novel (least visited) arm on any given trial, and normal rats alternate spontaneously at a rate of 70–80%. The effect of lead on spontaneous alternation has been studied.

Kostas et al. (1976) reported that rats dosed with low levels of lead showed a lowered rate of alternation, approaching random levels (50%). This was also noted by Jones (1979) in adult rats given 225 mg Pb/kg by gavage from postnatal days 1–20. Kostas *et al.* (1978) also reported that reduction in spontaneous alternation occurred after amphetamine administration in controls, but lead-dosed animals, which did not alternate after saline injection, paradoxically increased their percent alternation to around 80% when given amphetamine. DeRossett (1982) studied normal and amphetamine-induced spontaneous alternation in adult rats dosed with 0.5% lead acetate (2625 ppm Pb) for 2–4 weeks. In both non-drug and amphetamine trials percent alternation remained below significance between control and lead-dosed groups.

Other reports of changed or unchanged spontaneous alternation must be interpreted with caution. Spontaneous alternation is a non-stressful task in which the animal may freely explore its environment. Alternation tasks with components of reward, e.g. go/no go in a Skinner box (Hastings *et al.*, 1979) or alternation to tone presentation (Milar *et al.*, 1981) are generally regarded as non-spatial. The spontaneous alternation noted by Flynn *et al.* (1979) was carried out in a radial arm maze and may give some measure of alternation behaviour. The presence of cues from other arms, however, may merge the distinction between exploratory and alternation activity. Once more an effect of lead is not equivocal, but lower dose studies appear to show this paradigm is not sensitive to lead exposure.

To determine at what stage lead treatment produces its most severe effect on the development of exploratory activity in rats, Crofton *et al.* (1980) conducted cross-fostering experiments with the offspring of maternal rats that had been given 200 ppm lead chloride in their drinking water from 2 weeks prior to breeding until weaning at 21 days after parturition. This treatment produced mean blood lead concentrations in the maternal rats of 37.8 μg Pb/100 ml on day 18 of gestation and 35.5 μg Pb/100 ml at weaning. Some of the neonates were left with their mothers for 21 days from birth to weaning, while others were suckled by non-exposed maternal rats, and a third group born of non-exposed neonates were fostered by exposed maternal rats. Mean blood lead levels in these groups of neonates at weaning were 29.9, 14.5 and 36.0 μg Pb/100 ml respectively compared with 4.8 μg Pb/100 ml for the control group. Bodyweight, food and water consumption were unaffected in both maternal rats and neonates. However, in neonates exposed *in utero* and during lactation, the development of exploratory and locomotor activity (assessed at 5–21 days of age) was significantly delayed, by roughly 1 day from day 16 onwards. Litters treated only prenatally with lead showed the same delay, whereas those exposed only postnatally did not differ from unexposed controls. Differences in levels of activity followed the same pattern.

LEARNING PARADIGMS

Table 7b shows the large number of studies devoted to the effects of lead on learning (or more formally, conditioned behaviour), from simple learning through to the more complex responses to operant reinforcement. Learning is therefore a global term and different skills and/or knowledge will be used by the animal in different tasks, as and if they are required. The possible effects of lead on the ability of an animal to learn or remember any given task may be dependent on intrinsic (e.g. level of arousal, stress) or extrinsic (e.g. complexity) factors. All these must be taken into consideration. Discrimination paradigms have been used to study the effects of lead on learning behaviour. In these, the animal must learn to respond to one particular cue. This task may be considered difficult depending upon the differences between cue choices. The first discrimination task was that reported by Brown (1975). The behavioural task employed was a successive light on/light off discrimination conducted in a T-maze, and a poorer performance was observed in lead-dosed rats. This was considered a relatively difficult task and Brown suggested that lead was causing a change in hippocampal function, although this task is not especially sensitive in animals with hippocampal damage. Other visual discrimination studies followed.

Impairment was reported in lead-dosed rats in black/white (Brady et al., 1975); light on/light off (Driscoll and Stegner, 1976); in difficult (circle size) but not easy (horizontal/vertical stripes) tasks on the Lashley jumping stand (Winneke et al., 1977; 1982); on both brightness and shape (Zenick et al., 1978); spatial but not light (Lanthorn and Isaacson, 1978); bright/dim but not successive (Hastings et al., 1977; 1979) discrimination. Impairment in visual discrimination has also been shown in other species, notably pigeons (Barthlamus et al., 1977), sheep (Carson et al., 1974), monkeys (Rice and Willes, 1979) and rhesus monkeys (Bushnell and Bowman, 1979).

Some of these studies deserve further detailed examination. The effects of lead exposure from the prenatal period onwards were assessed by Winneke et al. (1977), who fed female rats a diet containing 1380 ppm lead acetate (754 ppm Pb) from 60 days before mating until weaning and fed their offspring on the same diet after weaning. Average blood lead levels in the mothers were 24.2 μg Pb/100 ml before mating and 31.2 μg Pb/100 ml after weaning, whereas in their offspring, levels were 26.6 and 28.5 μg Pb/100 ml at 16 and 190 days of age respectively. The level of lead given was sufficient to reduce the proportion of animals that became pregnant and the size of litters, but the offspring were heavier than the controls (probably because of the reduced litter size). Between 100 and 200 days of age, a total of 40 male offspring were tested in two visual discrimination learning tasks. In the 'easy' task there was no difference from controls, but in the 'difficult' task only one of ten treated animals learned the problem within 27 days, compared with eight of ten controls. The rats were also subjected to an open field test, in which those given lead showed an increase in ambulation, rearing and grooming.

Hastings et al. (1979) showed that the learning ability of adult rats was impaired by neonatal treatment with lead, despite a subsequent absence of abnormal exposure. Lactating rats were given 0.02 or 0.2% lead acetate

(109 or 1090 ppm Pb) in their drinking water for 21 days, treatments which produced average milk levels of 21 and 139 μg Pb/100 ml and blood lead levels in the offspring of 29 and 65 μg Pb/100 ml respectively, compared with 11 μg/100 ml in the blood of controls. Brain lead levels were similar to those in blood. The offspring were then weaned onto tap water (<0.05 ppm Pb) and normal diet. At 120 days of age the low lead animals took about 50% longer than the controls, whose mothers had been given either tap water or sodium acetate, to learn a simultaneous visual discrimination task, while the high lead group took even longer. However, at 270 and 330 days, when the same rats were also treated in a successive visual discrimination task and in a go/no-go discrimination task, no significant differences from controls in either of these tests were found.

Geist and Mattes (1979) suggested that lead at levels of 50 ppm and even less in the drinking water may affect the behaviour of rats, without producing obvious signs of toxicity. Rats given 0, 25 or 50 ppm lead as the acetate for 35 days, beginning at 23 days of age, showed no stunting of growth or weight loss, but both treated groups displayed impaired learning ability when tested in a Hebbs–Williams closed-field maze (Geist and Mattes, 1979). The Hebbs–Williams maze represents a rather sophisticated series of problems which have been used in lesion studies. Unfortunately blood lead levels were not measured in this study but as the animals were fed on a normal diet these were probably comparable with those reported by Cory-Slechta and Thompson (1979) (see below).

A number of studies of the effect of lead on performance in the 8 arm radial maze have been carried out. Flynn et al. (1979) used two dosing procedures: perinatal administration (2625 ppm Pb) and a three stage, longterm dosing regimen (1090 ppm Pb to maternal rats; 225 μg Pb to neonates and 1363 ppm Pb to weaned animals). No effect of lead was found in the number of arms entered or in a measure of spontaneous alternation (see above) in any dose group.

Operant conditioning experiments, in which the probability of a response in a particular stimulus environment is increased by following that response with reinforcement, have been very useful in behavioural studies. There are two types of schedule reinforcement: ratio and interval. In the ratio type, reinforcement occurs after a defined number of responses (e.g. bar presses); in interval, after a defined time. Some variation may be introduced by changing the schedules (within certain limits). These procedures are sensitive to changes in the stimulus environment (both internal and external), and thus provide a natural barometer for assessing the effects of lead. Learning tasks that use either positive or negative reinforcements seem equally sensitive to disruption following lead exposure. Therefore, it is unlikely that lead has a selective effect on any specific motivational system, and may in fact interfere directly with 'learning' processes.

Alterations in fixed ratio responding have been reported by Zenick et al.

(1977b) and Angell and Weiss (1982). The majority of studies have used fixed interval schedules, which appear more sensitive, as a degree of response control must be made by the animal. Increased response rates and greater intersubject variability were observed by Cory-Slechta and Thompson (1979) and a poorer performance overall was noted by Zenick *et al.* (1979b) in lead-dosed rats. The poorer performance noted in response inhibition reported by Overmann (1977) is an example of a fixed interval schedule, as is the alteration in temporally-spaced responding described by Dietz *et al.* (1978).

Only one of these studies gave blood lead levels. Cory-Slechta and Thompson (1979) administered drinking water containing 0, 50, 300 or 1000 ppm lead acetate from weaning at 20–22 days of age, and were fed simultaneously on a diet of unmeasured lead content resulting in blood lead levels at 150 days of age of approximately 5, 6, 26 and 42 μg Pb/100 ml respectively. After 35 days of exposure, behavioural training was begun in which they were assessed on their responses to a fixed interval food reinforcement schedule (involving a food reward for the first lever press response occurring at least 30 s after the preceding food delivery). Exposure to 50 and 300 ppm increased response rates (frequency of lever pressing) and intersubject variability, while latency, or time to first response in the interval, decreased. At 1000 ppm there was initially a decrease in response rates and an increase in latency values. Following termination of exposure at 50 ppm, response rates and latency values gradually returned to control levels. Behavioural effects produced by 50 or 300 ppm were similar in magnitude but varied in time to onset and decline. Why the lowest dose level used in this study affected behaviour, despite producing blood lead levels not significantly greater than those in control rats, is uncertain. As such levels were measured only at 150 days, it is possible that higher body levels were present earlier during the treatment procedure. This is possible since lead absorption from the gut is far greater in young than in adult animals. Alternatively, even 50 ppm lead might be toxic in these animals, producing a secondary effect on behaviour. (See e.g. Reiter *et al.*, 1975; Geist and Mattes, 1979).

Cory-Slechta *et al.* (1981) investigated the performance of rats (five/group) given 100 or 300 ppm lead acetate (54.6 or 163.9 ppm lead) in drinking water from weaning. At 55 days of age they were trained to press a lever for food reward, after which a schedule was imposed in which only lever-pressing for longer than a certain time was rewarded (fixed ratio). Lead-treated groups showed more variable responses than controls, but on average their response duration was shorter, and they paused for longer before responding. No dose–response relationship was apparent at the two levels used. Blood lead levels were not measured, but brain levels at 95 days of age were in the range 40–142 ng/g at 100 ppm and 320–1080 ng/g at 300 ppm (14–26 ng/g in controls).

Temporally-spaced responding is a modified fixed interval schedule, more commonly known as differential reinforcement of low (or high) rate. In this schedule, the animal must make its responses a specified period apart (e.g. DRL-20 requires a bar-press every 20 s). Apart from the deficits reported by Dietz *et al.* (1978), other studies have investigated the effects of lead on differential reinforcement. These include deficient performance (Alfano and Petit, 1981); increased response latency and decreased response duration (Cory-Slechta *et al.*, 1981) with low rates; and increased performance in rats with a blood lead of 20 μg Pb/100 ml, but decreased performance at a steady-state blood lead of 60 μg Pb/100 ml (Gross-Selbeck and Gross-Selbeck, 1980; 1981).

COMPLEX BEHAVIOUR

Reported findings are abstracted in Table 7c. Neonates from maternal rats that had been given 200 ppm lead acetate in the drinking water from parturition to weaning had mean blood lead levels of 21.7, 25.2 and 2.5 μg Pb/100 ml at 10, 21 and 60 days of age respectively, while brain levels at these stages were 6.3, 12.5 and 6.9 μg Pb/100 ml. In controls, lead levels at 60 days were 3.0 μg Pb/100 ml in blood and 3.2 μg Pb/100 ml in brain. The lead-treated neonates showed no effects on bodyweight, but the time to eye opening was significantly delayed, confirming the findings of Reiter *et al.* (1975). The time taken by the neonates to right themselves when placed head downwards on a slope was increased at the 10% but not the 5% level of significance. Subjecting the neonates to a maximal electroshock seizure test decreased the ability of lead-treated 10–14-day-old neonates to maintain suspension from a taught wire by forepaw grip, but did not affect the appearance of the auditory startle response or the air righting reflex of lead-treated neonates (Overmann *et al.*, 1977). Lead also caused more severe seizures in response to electroshock treatment by day 16, and this increased severity of response was still evident at day 60, 39 days after cessation of exposure (Fox *et al.*, 1979b).

DRUG-INDUCED BEHAVIOUR

The production of artificial behaviour by pharmacological challenge is a useful method for exploring the functional consequences of relatively subtle alterations in CNS mechanisms. Stimulation or inhibition of neurotransmitter function may evoke patterns in behaviour that will not be normally seen. This may facilitate comprehension of neurological dysfunction or impairment. The administration of centrally acting agonists or antagonists to lead-dosed animals has been carried out in an attempt to uncover changes which may be obscured by the complex neurochemical interplay underlying normal behaviour; these are briefly reviewed in Table 7d.

Catecholaminergic agents

Amongst catecholaminergic agonists, the most frequently used drug has been amphetamine, which usually produces an increase in locomotor activity. Silbergeld and Goldberg (1974a and b) reported a paradoxical decrease in activity induced by L- and D-amphetamine in mice given drinking water loaded with 10 g/l lead acetate (5450 ppm) and showing undernutrition and hyperactivity. Sobotka and Cook (1974) reported that amphetamine-induced locomotor activity was attenuated in rats given 81 mg Pb/kg orally from postnatal days 3–21, and suffering from undernutrition but not hyperactivity. A diminished responsiveness to amphetamine was also reported by Reiter *et al.* (1975) in a longterm study (from 40 days preconception in maternal rats and throughout life in their offspring) in rats exposed to 5 and 50 ppm Pb and showing no weight gain differences and reduced activity. Memo *et al.* (1980a) reported that rats given 2.5 g Pb/l (1365 ppm Pb) exhibited hyperactivity, and showed decreased amphetamine-induced activity.

In each of these studies amphetamine-induced locomotor activity in lead-dosed animals has been reported to be either (a) attenuated, i.e. diminished responsiveness to the drug, or (b) paradoxically decreased. This alteration is independent of any initial activity state.

Other dopamine agonists have been used. In a neuropharmacological study in mice, Silbergeld and Goldberg (1975) reported a number of catecholaminergic changes in mice dosed with 5 g/l lead acetate (2731 ppm Pb), and showing undernutrition and hyperactivity. Enhancement of catecholaminergic function was achieved with L-DOPA (a precursor), benztropine (which inhibits catecholamine uptake) and apomorphine (a direct receptor agonist). These drugs all caused an exacerbation of existing hyperactivity in lead-dosed animals with minimal or no effects in controls. The paradoxical decrease of amphetamine-induced activity was reproduced with both amphetamine and methylphenidate (a structurally related drug). This was also seen with fenfluramine (a phenylethylamine known to affect aminergic pathways). The tyrosine hydroxylase inhibitor α-methylparatyrosine greatly reduced motor activity in lead-dosed animals, while controls were not significantly affected. This treatment also appeared to reverse the effect of amphetamine, so that lead-treated animals responded like naive controls.

Antagonists of catecholaminergic function have also been used. The neuroleptics haloperidol (a centrally acting butyrophenone) and (–)sulpiride have been administered to lead-dosed animals. Lucchi *et al.* (1981) found that rats dosed with 2.5 g Pb/l (1365 ppm Pb) showed no differences in haloperidol-induced sedation, while the dose of (–)sulpiride which caused sedation was lower in lead-intoxicated animals than in control rats. These observations suggest that one of the neurochemical changes that may be ascribed to lead is an alteration in receptors sensitive to (–)sulpiride, i.e. a discrete population of dopaminergic D2 receptors.

In each of these studies, catecholaminergic function has been shown to be altered in some manner. However, the reported somatic or nutritional changes make assessment of these results difficult. Michaelson (1980) has evaluated the effects of various levels of undernutrition on later levels of locomotor activity and response to amphetamine. A statistically significant relationship was found between bodyweight and change in activity both before and after injection. It is evident that undernutrition and not lead as such, may be at least partially responsible for some of the behavioural effects previously reported. The experiments reported in chapter 3 have shown that the effects of undernutrition (an observable index of indirect toxicity) becomes evident at dose levels of about 1000 ppm Pb in drinking water. Changes in activity levels have been reported in undernourished animals (Loch *et al.*, 1978; Zenick *et al.*, 1979c), so that findings of studies employing doses above this level must be interpreted with caution.

The problems of overt toxicity and subtoxic lead exposures have been addressed in other studies. Wince *et al.* (1980) used the offspring from maternal rats who had received 4% lead carbonate in the diet and had been weaned to 40 ppm Pb, but with pair-fed animals to control for the effects of undernutrition. Lead-dosed animals were significantly more active than controls, but there were no statistical differences in amphetamine- and apomorphine-induced motility between groups. These results suggest that the previously observed effects of catecholaminergic agonists using high levels of lead exposure were possibly due to indirect lead-induced changes.

In a low-dose study, rats receiving 25 and 75 mg/kg lead acetate by mouth from postnatal days 2–14, had blood leads at day 15 of $50 \mu g$ and $100 \mu g$ Pb/100 ml respectively (Jason and Kellogg, 1981). There were no somatic differences between lead-dosed and control animals up to 35 days, when the experiment was terminated. At this age there were no significant differences in activity between groups. Activity scores fell after a predrug injection of saline, possibly due to habituation of animals to the apparatus. The low-dosed groups had significantly lower activity than controls or high dose groups. This phenomenon was also evident and statistically significant after amphetamine injection at a dose of 1 mg/kg. This effect of lead was not significant after a high dose of amphetamine (5.0 mg/kg), although qualitative assessment of activity showed that low-dose animals exhibited more stereotyped behaviour than the other two groups. These results are difficult to interpret, but suggest that lead exposure at low doses might be causing a direct alteration in amphetamine-mediated mechanisms.

Support for this idea comes from Rafales *et al.* (1981). Their study reported no change in activity in rats dosed with 109 ppm Pb postnatally. The possibility of observable general toxicity at this dose is extremely unlikely and blood lead levels were reported in the 'normal' human range ($20–30 \mu g$ Pb/100 ml). However, amphetamine administration to male rats caused a significant (21%) increase in locomotor activity over that of controls.

Cholinergic agents

Silbergeld and Goldberg (1974a; b) studied the effects of lead-induced hyperactivity in the cholinergic system (Silbergeld and Goldberg, 1975). Agonism was investigated with physostigmine (which inhibits the acetylcholine degrading enzyme acetylcholinesterase), and with dimethylaminoethanol (a proposed precursor of choline and a drug which increases brain acetylcholine). Physostigmine produced no changes in the activity of control mice but reduced motor activity of lead-dosed animals; dimethylaminoethanol also suppressed hyperactivity in lead-treated animals, while observable behavioural excitation and stimulation of motor activity were seen in controls. Possible peripheral involvement was studied with neostigmine (a quaternary acetylcholinesterase inhibitor that has little penetration of the blood–brain barrier). Neostigmine had no effect on lead-induced hyperactivity.

Inhibition of cholinergic function was studied by blockade of muscarinic receptors with atropine and benztropine (drugs known to inhibit the high affinity transport of choline into synaptosomes). Lead-treated hyperactive mice became hyperactive to stimuli following atropine and showed increased activity levels following benztropine (Silbergeld and Goldberg, 1975).

Other pharmacological agents

Modification of neurotransmitter systems involving GABA may be associated with changes in CNS excitability, producing seizures. Evaluation of pharmacological challenge with GABAergic agents will therefore include measurement of seizures and convulsion susceptibility.

Silbergeld et al. (1979) described various measures of increased seizure activity in rats treated with very high doses of lead. The intravenous infusion of picrotoxin (a powerful CNS stimulant), isoniazid (a pyridoxal phosphate antagonizing hydrazine derivative that produces convulsions) and strychnine (an alkaloid stimulant) caused convulsions in lead-dosed rats at lower doses than in controls. The intraperitoneal injection of 3-mercaptopropionic acid caused clonus in a shorter time in lead-dosed animals than in controls. Lead exposure did not affect the responses to pentylenetetrazol (an analeptic). There is evidence that seizure susceptibility is itself affected by exposure to lead (Fox et al., 1979b; Burright et al., 1982). However, these results suggest the possibility that the neurochemical function of GABA-containing neurones might be altered following lead treatment.

CNS sedatives have also been administered to lead-dosed animals. Phenobarbitol was reported to exacerbate the behaviour of lead-treated mice (Silbergeld and Goldberg, 1974a and b), chloral hydrate sedation was reported to be unchanged in lead dosed mice (Silbergeld and Goldberg, 1974a), although chlorpromazine-induced sedation appeared to be altered

(Silbergeld and Goldberg, 1975). In the latter study the onset of sedation and duration was shortened as compared with controls.

Morphine-induced catalepsy has also been studied. The response to morphine may be of two types, since both reduction of activity with enhancement of cataleptic response and increase in hyperactivity and decrease in response to morphine challenge have been noted (Mitchell *et al.*, 1982). These results are confusing but do not exclude the possibility of changes in opioid-induced behaviour.

There have been some studies attempting to link normal or drug-induced behaviour with brain neurochemistry. As yet these studies (Memo *et al.*, 1980a; Rafales *et al.*, 1981) have largely failed to relate behavioural changes to altered neurochemistry, although one report (Lucchi *et al.*, 1981) has made limited correlation.

SUMMARY

In summary, a review of the effects of lead on behaviour in the rat shows great diversity of findings. Only recently has adequate simulation of the exposure conditions found in young children been made. The measurement of blood lead makes correlation between studies easier, and is a useful index of exposure, helping to clarify the large number of regimens of lead administration that have been reported. There may be more accurate ways of assessing lead load, but blood lead measurements reflect a balance between convenience and accuracy. Other indices, such as dose or equivalence of effect may be less reliable and may make comparison between different studies difficult.

The selection of behavioural tasks is often based on what appear to be historic or logistical precepts, and tends to be made with little or no apparent regard to the formulation or testing of hypotheses. With respect to general experimental design, several tactical deficiencies frequently occur. These include small sample sizes; inappropriate application of statistical methods; or the presence of confounding variables (such as undernutrition). The replication of some experiments, often lacking, would lend credibility to their findings. It is possible to conclude that failure to observe a change in task acquisition or performance may occur because the task is not sensitive enough, or that the neural systems that mediate the behaviour under test are not affected by lead at the exposure levels examined.

Activity studies give the most variable results, but there is the suggestion that exposure to lead causes an increase in behavioural reactivity. Activity measurements are a 'global' representation of brain function, and therefore altered activity levels are at best a non-specific indication of nervous system dysfunction. There is the suggestion that increased locomotor activity in young animals occurs at moderately high exposures of lead, but this tendency is absent at lower levels. If anything, longer term, low level exposure seems to produce hypoactivity.

It is not possible to define experimentally the effects of lead on cognitive function (e.g. learning and memory), and effects on arousal and motivation can complicate the observed picture. The definition of acquisition may also represent performance, as is the case in studies that report group differences on the first day of acquisition. Some types of learning problems are more sensitive to lead-related disruption than others. Simple pattern discrimination and passive avoidance are not especially sensitive to lead. Functional impairment by lead appears to increase as the task complexity increases (e.g. from discrimination tasks to differential reinforcement of low rate operant responding). This suggests that the possible consequences of lead exposure will only be uncovered when experimental design becomes more sophisticated than at present.

The effect of lead on complex behaviour, such as social or aggressive behaviour, has been investigated in a relatively small number of studies. This category of experiment partly reproduces activity measurement in that the animal is observed for behavioural change with no experimental manipulation. There have been some reports of increased aggression and altered sexual behaviour; these are doubtful. The reported changes to electroshock in lead-dosed animals appear more specific, but functionally uninterpretable.

The results of experiments using pharmacological challenge are more promising. By stimulating or inhibiting individual neurotransmitter systems, a neurochemical basis for a functional change can be established. Problems of non-specificity may again occur, and only when these are overcome can experiments reveal direct effects. To be sought are effects on catecholaminergic and GABAergic systems, and possibly on peptidergic activity.

5
The Neurochemical Effects of Lead

Although it is possible to measure certain neurochemical parameters in humans, the understanding of molecular mechanisms of lead toxicity must depend on animal experimentation. Alterations in neurochemistry may precede other neurotoxic effects and be the basis for the expressions of altered behaviour, pharmacological response or cell pathology.

EXPERIMENTAL APPROACHES

Before reviewing the reported effects of lead on neurochemistry, it is necessary to discuss the more general effects of lead on brain biochemistry. A great deal of work has been carried out on the possible general metabolic effects of lead, but for the purposes of this chapter, only biochemical effects on the brain will be discussed.

BIOCHEMICAL EVIDENCE

Table 8 shows the reported biochemical effects of lead on the CNS. As many of these studies used very high doses of lead, it is likely that some of the reported lead-induced changes represent non-specific effects. Prominent in this category are the changes in protein of DNA concentrations and decreases in respiratory activity.

The first general biochemical study of the effects of lead on cerebral cortex and cerebellum, using 4% lead carbonate in the maternal diet from birth, was made by Michaelson (1973). He found a transient postnatal growth retardation and reduction of cerebral and cerebellar wet weights. The cerebellum showed increased water and reduced DNA content, suggesting a 15–20% deficit in cell numbers. DNA in the cerebrum and RNA and protein in both areas differed only marginally from controls, although all were consistently

Table 8 Biochemical effects of developmental lead exposure in rats and mice

Biochemical parameter	Area	Age	Effect	Reference
Protein	Cortex		No change	Michaelson, 1973
	Cerebellum		No change	Michaelson, 1973
	Whole brain		No change	Gerber *et al.*, 1978
			Decreased above 3000 ppm	Srivastava and Thakur, 1981
RNA	Cortex		No change	Michaelson, 1973
	Cerebellum		No change	Michaelson, 1973
			Decreased above 3000 ppm	Srivastava and Thakur, 1981
DNA	Cortex		No change	Michaelson, 1973
	Cerebellum	3 weeks	Decreased	Michaelson, 1973
	Whole brain		Increased	Gerber *et al.*, 1978
Development of brain DNA			Retarded	Gerber *et al.*, 1978
			Decreased	Srivastava and Thakur, 1981
Amino acids				
Tyrosine				
endogenous levels	Whole brain		No change	Schumann, 1977
uptake	Forebrain	40–90 days	Increased	Silbergeld and Goldberg, 1975
transport	Capillary	55–70 days	No change	Michaelson and Bradbury, 1982
Phenylalanine				
uptake	Forebrain	40–90 days	No change	Silbergeld and Goldberg, 1975
Glycine				
uptake	Forebrain	40–90 days	No change	Silbergeld and Goldberg, 1975
levels	Plasma		Decreased	Hsu, 1981
Leucine				
uptake	Forebrain	40–90 days	No change	Silbergeld and Goldberg, 1975
Histidine				
levels	Plasma		Decreased	Hsu, 1981
Glutamate				
levels	Plasma		Decreased	Hsu, 1981

Table 8 *continued*

Biochemical parameter	Area	Age	Effect	Reference
Serine				
levels	Plasma		Decreased	Hsu, 1981
Proline				
levels	Whole brain		Increased	Gerber *et al.*, 1978
Developmental increase				
glutamate	Cortex		Decreased	Patel *et al.*, 1974a
	Cerebellum		Decreased	Patel *et al.*, 1974a
aspartate	Cortex		Decreased	Patel *et al.*, 1974a
	Cerebellum		Decreased	Patel *et al.*, 1974a
glutamate	Cortex		Increased	Patel *et al.*, 1974a
	Cerebellum		No change	Patel *et al.*, 1974a
Compartmentation			Delayed	Patel *et al.*, 1974b
Glycolysis				
Glucose-6-phosphate dehydrogenase		day 20	Decreased at high dose	Gelman *et al.*, 1979
6-Phosphogluconate dehydrogenase		day 6, 20	Decreased at high dose	Gelman *et al.*, 1979
Tricarboxylic acid cycle				
development	Cortex		Retarded	Patel *et al.*, 1974a
	Cerebellum		Retarded	Patel *et al.*, 1974a
flux	Cortex		Depressed	Patel *et al.*, 1974b
	Cerebellum		Depressed	Patel *et al.*, 1974b
Mitochondrial				
Phosphorylation			Inhibited	Krall *et al.*, 1972
Respiratory control ratio		28 days	Decreased	Holtzman and Hsu, 1976
				Holtzman *et al.*, 1980
NAD-linked ADP/O ratio		28 days	Decreased	Holtzman and Hsu, 1976
NAD-linked phosphorulation coupled respiration			Inhibited	Holtzman *et al.*, 1980
ADP-independent respiration		21 days	Increased	Holtzman and Hsu, 1976
ADP-dependant respiration		28 days	Decreased	Holtzman and Hsu, 1976

Table 8 *continued*

Biochemical parameter	Area	Age	Effect	Reference
Cytochromes				
Cytochrome oxidase activity	Cortex	28 days	Increased	Holtzman and Hsu, 1976
			Decreased	Gatzke, 1980
accumulation	Cortex	10, 15 days	Delayed	Bull *et al.*, 1979
levels				
Cytochrome b	Cerebellum	28 days	Decreased	Holtzman *et al.*, 1981
c + c1	Cerebellum	28 days	Decreased	Holtzman *et al.*, 1981
aa3	Cerebellum	28 days	Decreased	Holtzman *et al.*, 1981
	Cortex	21 days	Decreased	Holtzman *et al.*, 1981
P-450	Cortex	14 days	Increased	Holtzman *et al.*, 1981
b5	Cortex	14 days	Increased	Holtzman *et al.*, 1981
Aminolaevulinic acid dehydratase	Cerebellum	15, 28 days	Increased	Holtzman and Hsu, 1976
			Decreased	Barlow *et al.*, 1977
			Decreased	Gerber *et al.*, 1978
Glutathione metabolism			Increased	Hsu, 1981
Glutathione peroxidase		day 6, 13	Decreased at high dose	Gelman *et al.*, 1979
Lysosomal				
β-N-Acetylglucosaminidase release		3 weeks	Inhibited	Hsu, 1981
Acid phosphatase activity		3 weeks	No change	Hsu, 1981
			No change	Gerber *et al.*, 1978
			Decreased	Gatzke, 1980
Cathepsin			Increased	Gerber *et al.*, 1978
Catalase			No change	Gelman *et al.*, 1979
Alkaline phosphatase			Decreased	Gatzke, 1980
Alpha-naphthylesterase			No change	Gatzke, 1980
Lipid metabolism				
Neural lipids			Decreased	Krigman *et al.*, 1972
Sialic acid			Decreased	Gerber *et al.*, 1978
Phospholipids			No change	Gerber *et al.*, 1978
			No change	Stephens and Gerber, 1981

78

Table 8 *continued*

Biochemical parameter	Area	Age	Effect	Reference
Cerebrosides		32 days	Decreased at 0.1%	Stephens and Gerber, 1981
Sulphatides		32 days	Decreased at 0.1%	Stephens and Gerber, 1981
Gangliosides		32 days	Slight decrease	Stephens and Gerber, 1981
Sphingomyelin			No change	Stephens and Gerber, 1981
Cholesterol			No change	Stephens and Gerber, 1981
Lipid peroxidation				
Malondialdehyde formation			No change	Gelman and Michaelson, 1979
Lipofuscin pigment concentration			No change	Gelman et al., 1979
Superoxide dismutases				
mitochondrial			No change	Gelman et al., 1979
cytosolic			No change	Gelman et al., 1979

lowered. Michaelson speculated that perhaps interference with respiratory enzymes affected the metabolic capacity of the developing cerebellum, an idea taken up and investigated by Patel *et al.* (1974a and b).

Patel and colleagues also used the same dosing regime and studied metabolism of radiolabelled [^{14}C]glucose and [^{3}H]acetate in the cerebrum and cerebellum. Their results indicate that although uptake of [^{14}C]glucose was affected during chronic exposure, a decrease in transamination reactions reflected a reduction in the utilization of glucose by both regions (Patel *et al.*, 1974a). It was also observed that the conversion of [^{3}H]acetate into amino acids was decreased in the lead-treated rat. Both these results are consistent with a depression of Krebs cycle activity in the brain. A delay was seen in maturation of normal compartmentation as adduced from altered glutamate:glutamine ratios (Patel *et al.*, 1974b).

The idea of impaired energy metabolism was further investigated by Holtzman and Hsu (1976). They studied the respiration of isolated brain mitochondria, again using 4% lead carbonate. They observed two phases of lead effect, probably related to the increasing concentration of lead during the first few days. The early changes appeared within 2 days in the cerebellum and were seen as significantly lowered ADP-dependent respiration, due to an increase in ADP-independent respiration. During the next 2 weeks both ADP-dependent and ADP-independent respiration were progressively inhibited, as was the activity of cytochrome oxidase, the terminal enzyme of the respiratory chain. All these changes were also seen in the cerebral hemispheres but were smaller and statistically non-significant. Lead appears to act on mitochondrial respiratory enzymes like an uncoupling agent such as dinitrophenol, although with less efficiency. A decrease in glycolytic enzyme activity was reported by Gelman *et al.* (1979) using doses of 225 mg Pb/kg by gavage. This disruption of energy metabolism could well produce dramatic effects on the developing, metabolically active nervous system.

Other biochemical effects include changes in the lipid composition of the brain. Significant deficiencies of phospholipids, gangliosides and cholesterol were all evident in post-weaned lead-exposed animals, although these were not found in adult animals (Krigman *et al.*, 1972; 1974a; Krigman and Hogan, 1974). Stephens and Gerber (1981) amplified these measurements and suggested that the deficiencies observed reflected a selective disorder of cellular metabolism. Using a wide range of doses, from 500 ppm to 6000 ppm Pb, Srivasta and Thakur (1981) examined various biochemical effects. Changes were generally found above 1000 ppm Pb, and always above 3000 ppm Pb. Undernutrition and obvious toxicity accompanied decreases in RNA, DNA and protein concentration.

Lead has been shown to inhibit calcium competitively in a variety of biological systems, e.g. in the superior cervical ganglion of the cat (Kostial and Vouk, 1957); in presynaptic terminals in frog sympathetic ganglion (Kober and Cooper, 1977); in adrenergic synapses in the rabbit saphenous artery (Cooper and Steinberg, 1977); and in the cholinergic system (Carroll *et al.*, 1977; Silbergeld, 1977). The CNS interactions between lead and calcium are dependent on the neurotransmitters involved.

Low dose studies have been carried out, and possibly more specific effects noted. Disruption in cytochrome metabolism was observed by Vanderkooi and Landesberg (1977) who reported synthesis of iron-free cytochrome during lead intoxication. Two related studies by Bull *et al.* (1979) and McCauley *et al.* (1979), using much less toxic doses of lead (5, 30 and 200 ppm Pb), over a much longer period (14 days before breeding till weaning), showed a dose-dependent depression of cytochrome accumulation in post-natal cerebral cortex (Bull *et al.*, 1979). More evidence of uncoupling came from McCauley *et al.* (1979) who found increased oxygen and glucose consumption in response to an increased potassium concentration. This response represents an increased energy demand that is coupled to oxidative phosphorylation, although the increased response in lead-treated animals could conceivably have been due to an inefficient transport system for potassium. Uncoupling of energy metabolism has been implicated in delays in cerebral cortical development (McCauley *et al.*, 1979) and locomotor activity (Crofton *et al.*, 1980). Blood leads measured at 200 ppm Pb were 36 μg Pb/100 ml blood for weaned neonates, suggesting a consequence of lead exposure at a level not uncommon in human populations.

A wide variety of enzymes are known to be sensitive to lead exposure. An extensive description of the enzymatic effects of lead is reviewed in U.S. E.P.A. (1977). As with several other metals, lead has a high affinity for various complexing groups, such as the imidazole, cysteine sulphydryl and e-amino groups of lysine. An effect may be imparted by alteration to the structural integrity of enzymes, or by the disruption of substrate–enzyme binding.

Two enzyme systems have shown themselves to be extremely sensitive to lead at low levels. The first of these is D-amino laevulinic acid dehydratase (D-ALAD), the initial and rate limiting step in the porphyrin synthetic pathway. In both experimental (Barlow *et al.*, 1977) and clinical studies (Piomelli *et al.*, 1980) this enzyme is potently inhibited by lead. In a review of dose-dependent low level lead effects, Zielhuis (1975) calculated a 'no effect level' for this enzyme at about 10 μg Pb/100 ml blood in man. Other enzymes in the porphyrin synthetic pathway are also affected by lead, e.g. ferrochelatase, the enzyme responsible for the insertion of haem into the porphyrin precursor protoporphyrin IX (Moore, 1975), but this is probably due to D-ALAD-related interactions. Silbergeld and Lamon (1980) have speculated that the neurotoxic effects of lead may be due in part to a competitive interaction involving amino laevulinic acid at neuronal receptors, as there may be similarities between features of lead toxicity and some of the porphyrinopathic diseases (Moore *et al.*, 1980).

The second enzyme system that appears to be sensitive to low levels of lead

81

is that of biopterin metabolism. Biopterin (4,5,7,8-tetrahydrobiopterin) is a cofactor for both the hydroxylation of phenylalanine to tyrosine, and for the hydroxylation of tyrosine to the catecholamine intermediate L-3,4 dihydroxy-phenylalanine (L-DOPA). Purdy *et al.* (1981) reported that lead concentrations as low as 0.01 μmol/l produced a 35% inhibition in the *in vitro* synthesis of biopterin, decreasing to 55% at 10 μmol/l. The salvage pathway which regenerated oxidized biopterin was affected at lead concentrations above 1 μmol/l due to irreversible inhibition of dihydrobiopterin reductases. This is not confined to *in vitro* systems. Serum biopterin derivative levels have been positively correlated with blood lead levels in human patients (Leeming and Blair, 1980); and experimentally, *in vivo* interaction between lead and tetrahydrobiopterin metabolism has been shown. Male rats receiving 5 mg Pb/l in their drinking water (5 ppm Pb) since weaning showed a significant positive correlation between blood lead and plasma dihydrobiopterin after 3 months, and a fall in plasma dihydrobiopterin after 7 months (McIntosh *et al.*, 1982).

In vitro studies have been carried out on sodium- and potassium-dependent ATPase. Selhi and White (1975) suggested that conformational changes and alteration in spatial arrangement of proteins in red cell membranes were responsible for the inhibition of the sodium/potassium ATPase by lead at physiological concentrations. This effect was shown to be a non-competitive reversible reaction and the affinity of lead for brain ATPase was high (Seigel *et al.*, 1977). This was confirmed by Neuchay and Saunders (1978) who also showed that the site of action on the ATPase appeared to be the sodium-dependent phosphorylation site, and that ATP chelates lead. These observations have suggested a possible biochemical basis for the oedema of lead encephalopathy. Goldstein *et al.* (1974) showed that sodium concentration in brains developing encephalopathy increased by 88% and potassium only decreased (by 15%) in the later stages of oedema and haemorrhage. If lead were to induce failure in the sodium side of the sodium/potassium ATPase, then the cell poisoned by lead would be unable to pump out the sodium which normally diffuses in via Donnan equilibrium.

Nathanson and Bloom (1975) showed that brain adenylate cyclase is inhibited *in vitro* by lead at very low concentrations. An IC50 value was obtained of 3 μmol/l (equivalent to 62 μg/100 ml or 0.62 ppm). Inhibition might take place in the brain of normal rats, since Sauerhoff and Michaelson (1973) have shown brain lead concentrations of 0.1 ppm. These authors also reported that lead-exposed rats have brain leads of 0.6–1.2 ppm which caused inhibition of adenylate cyclase of above 50% in the parallel *in vitro* system.

NEUROCHEMICAL EVIDENCE

Many of the experimental studies of lead neurotoxicity have investigated effects of lead on specific neurotransmitters. Three neurotransmitters systems have been studied in detail, acetylcholine, catecholamines and γ-amino-butyric acid, although 5-hydroxytryptaminergic and more recently pep-tidergic systems have also been examined. This restriction of investigation should not be interpreted as implying that these systems are either exclusive or sensitive loci for the toxic effects of lead. Historical and logical reasons may account for the relative emphasis of research interest in these areas. The available evidence for the neurochemical effect of lead in rats and mice is summarized in Table 9.

Cholinergic effects

In animals subjected to early lead exposure, the steady state levels of acetyl-choline and its precursor choline were reported as unchanged in mouse fore-brain (Silbergeld and Goldberg, 1975; Carroll et al., 1977) and in the cerebellum, hippocampus, midbrain, medulla-pons and striatum of the rat (Modak et al., 1975). In the diencephalon, a small (20%) but significant increase was found in acetylcholine levels. The dosing regime used drinking water concentration of 1% (10 900 ppm Pb) which caused a 27% reduction in body weight at 60 days and a blood lead of 245 μg Pb/100 ml. No reliable change in acetylcholine levels were found in rat cortex by Shih and Hanin (1977; 1978a), but Hrdina et al. (1976) reported a 32–48% increase in treated rats.

The steady state levels of choline were reported to be reduced in the midbrain but not in other areas from rats exposed to lead via milk from maternal rats consuming a diet of 4% lead carbonate and weaned to a diet containing 40 ppm Pb (Shih and Hanin, 1977; 1978a).

Studies of synaptosomal high affinity binding transport of choline show 50% inhibition in one (Silbergeld and Goldberg, 1975) but no change in another (Carroll et al., 1977), though the same dose of lead was used in both studies (5 mg Pb/l – 2625 ppm Pb in the drinking water). A later study using administration of 0.1 mg Pb/kg by gavage, producing a blood lead of 402 μg Pb/100 ml also showed no effect on choline uptake (Ramsay et al., 1980). Carroll et al. (1977) were able to show an increase in spontaneous release of acetylcholine but not choline, although potassium-induced release of both was significantly lowered in lead-treated mice.

The effects of lead on endogenous enzymes involved in cholinergic metabolism are also uncertain. Sobotka et al. (1974) showed decreases in activity of acetylcholinesterase, while Carroll et al. (1977) found that acetylcholinesterase, choline acetyltransferase and choline phosphokinase were not altered by lead. However, Modak et al. (1975) demonstrated differ-ential effects of acetylcholinesterase and choline acetyltransferase in discrete areas of brain; lead treatment changed these enzymes in opposite directions. Acetylcholinesterase had significantly lower activity in the diencephalon,

Table 9 Neurochemical effects of developmental lead exposure in rats and mice

Neurotransmitter	Area	Age	Effect	Reference
CHOLINERGIC				
Acetylcholine				
endogenous levels	Forebrain	40–90 days	No change	Silbergeld and Goldberg, 1975
	Diencephelon	day 60	Increased	Modak et al., 1975
	Whole brain	21, 30 days	Decreased	Modak et al., 1978
	Whole brain	60 days	No change	Modak et al., 1978
	Cerebellum	30 days	Decreased at 1%	Modak et al., 1978
	Medulla	30 days	Reduced at 0.25, 0.5%	Modak et al., 1978
	Midbrain	30 days	Decreased at 0.5, 1%	Modak et al., 1978
		44–51 days	No change	Shih and Hanin, 1977; 1978a
	Diencephalon	30 days	Decreased at 1%	Modak et al., 1978
	Hippocampus	30 days	Increased at 1%	Modak et al., 1978
		44–51 days	No change	Shih and Hanin, 1977; 1978a
	Striatum	30 days	Decreased	Modak et al., 1978
		44–51 days	No change	Shih and Hanin, 1977; 1978a
	Cerebral cortex	30 days	Decreased at 0.25%	Modak et al., 1978
	Cortex	44–51 days	No change	Shih and Hanin, 1977; 1978a
	Half brain		No change	Carroll et al., 1977
turnover	Cortex	44–51 days	35% decreased	Shih and Hanin, 1977; 1978a
	Hippocampus	44–51 days	54% decreased	Shih and Hanin, 1977; 1978a
	Midbrain	44–51 days	51% decreased	Shih and Hanin, 1977; 1978a
	Striatum	44–51 days	33% decreased	Shih and Hanin, 1977; 1978a
KCl release	Half brain		Decreased	Carroll et al., 1977
Choline				
endogenous levels	Cortex	44–51 days	No change	Shih and Hanin, 1978a
	Hippocampus	44–51 days	No change	Shih and Hanin, 1978a
	Midbrain	44–51 days	30% decreased	Shih and Hanin, 1978a
	Striatum	44–51 days	No change	Shih and Hanin, 1978a
	Half brain		No change	Carroll et al., 1977
uptake	Forebrain	40–90 days	Decreased	Silbergeld and Goldberg, 1975
			No change	Ramsay et al., 1980
transport	Capillary	55–70 days	No change	Michaelson and Bradbury, 1982

Table 9 *continued*

Neurotransmitter	Area	Age	Effect	Reference
KCl release	Half brain		Decreased at 5, 10 mg/ml	Carroll et al., 1977
Acetylcholinesterase activity	Half brain	10 days	No change	Carroll et al., 1977
	Cerebrum	10 days	No change	Louis-Ferdinand et al., 1978
	Cerebellum	10 days	No change	Louis-Ferdinand et al., 1978
	Hippocampus	10 days	40% decreased	Louis-Ferdinand et al., 1978
	Midbrain	10 days	No change	Louis-Ferdinand et al., 1978
	Medulla	10 days	17% decreased	Louis-Ferdinand et al., 1978
	Striatum	10 days	No change	Louis-Ferdinand et al., 1978
	All areas	20 days	No change	Louis-Ferdinand et al., 1978
			Decreased	Gatzke, 1980
Cholinesterase activity	Cerebellum	60 days	No change	Modak et al., 1975
	Medulla	60 days	Decreased	Modak et al., 1975
	Midbrain	60 days	Decreased	Modak et al., 1975
	Diencephalon	60 days	Decreased	Modak et al., 1975
	Hippocampus	60 days	No change	Modak et al., 1975
	Striatum	60 days	No change	Modak et al., 1975
	Cortex	60 days	No change	Modak et al., 1975
	Pineal	60 days	No change	Modak et al., 1975
Butyrylcholinesterase activity	Cerebrum	20 days	Decreased	Louis-Ferdinand et al., 1978
	Hippocampus	15, 20 days	Decreased	Louis-Ferdinand et al., 1978
	Midbrain	15, 20 days	Decreased	Louis-Ferdinand et al., 1978
Choline acetyltransferase: activity	Cerebellum	60 days	No change	Modak et al., 1975
	Medulla	60 days	Increased	Modak et al., 1975
	Diencephalon	60 days	No change	Modak et al., 1975
	Hippocampus	60 days	Increased	Modak et al., 1975
	Striatum	60 days	No change	Modak et al., 1975
	Cortex	60 days	Increased	Modak et al., 1975

Table 9 *continued*

Neurotransmitter	Area	Age	Effect	Reference
Choline phosphokinase activity	Half brain		No change	Carroll et al., 1977
	Half brain		No change	Carroll et al., 1977
CATECHOLAMINES **Dopamine** endogenous levels	Half brain	day 21, 29	20% decreased	Sauerhoff and Michaelson, 1973
			No change	Michaelson et al., 1974
	Forebrain		No change	Golter and Michaelson, 1975
	Whole brain		No change	Silbergeld and Goldberg, 1975
	Striatum	day 15	20% decreased	Schumann, 1977
		8 weeks	No change	Jason and Kellogg, 1977
			Increased at 20, 40 ppm	Dubas and Hrdina et al., 1978
		day 15, 35	Decreased	Dubas et al., 1978
		100 days	No change	Jason and Kellogg, 1981
	Cortex	8 weeks	Decreased	Rafales et al., 1981
			Decreased	Dubas and Hrdina, 1978
	Midbrain	8 weeks	Decreased	Dubas and Hrdina, 1978
	Hypothalamus	8 weeks	Decreased	Dubas et al., 1978
			Increased at 40 ppm	Dubas et al., 1978
synthesis	Whole brain		No change	Schumann, 1977
	Forebrain	25–35 days	No change	Wince and Azzarro, 1977, 1978
			No change	Wince et al., 1980
uptake	Forebrain	40–90 days	Decreased	Silbergeld and Goldberg, 1975
	Forebrain	25–35 days	Decreased	Wince and Azzarro, 1977; 1978
			No change	Ramsay et al., 1980
			No change	Wince et al., 1980
turnover	Striatum	day 15	Decreased	Jason and Kellogg, 1981
	Half brain	day 32	No change	Michaelson et al., 1974
	Striatum	5 weeks	Slower decline	Govoni et al., 1978

Table 9 *continued*

Neurotransmitter	Area	Age	Effect	Reference
KCl release	Forebrain	25–35 days	Decreased	Wince et al., 1980
D-AMP release	Forebrain	25–35 days	No change	Wince et al., 1980
(−)Sulpiride stereospecific binding	Striatum	6 weeks	Increased	Lucchi et al., 1981
	Nucleus accumbens	6 weeks	Decreased	Lucchi et al., 1981
Monoamine oxidase activity	Cerebellum	6–8 weeks	Decreased	Krall et al., 1972
	Brain		No change	Cramer et al., 1980
Amphetamine uptake	Striatum	16–17 weeks	Decreased at 0.2%	Zenick et al., 1982
	Hypothalamus	16–17 weeks	Decreased at 0.2%	Zenick et al., 1982
	Hippocampus	16–17 weeks	Decreased at 0.2%	Zenick et al., 1982
	Midbrain	16–17 weeks	No change	Zenick et al., 1982
DOPAC levels	Striatum	5 weeks	20% decreased	Govoni et al., 1978
			Decreased	Govoni et al., 1979; 1980
			Decreased	Govoni et al., 1980
		6 weeks	Decreased	Memo et al., 1980a
		30 days after dosing	No change	Memo et al., 1981
	Nucleus accumbens	5 weeks	Increased	Govoni et al., 1979; 1980
			Increased	Govoni et al., 1980
		6 weeks	Increased	Memo et al., 1980a
		30 days after dosing	No change	Memo et al., 1981
	Substantia nigra	5 weeks	No change	Govoni et al., 1979; 1980
			No change	Govoni et al., 1980
	Frontal cortex	5 weeks	No change	Govoni et al., 1979
			Increased	Govoni et al., 1980
HVA levels	Striatum	5 weeks	20% decrease	Govoni et al., 1979
VMA levels	Forebrain	40–70 days	33% increased	Silbergeld and Chisholm, 1976
	Forebrain	40–70 days	48% increased	Silbergeld and Chisholm, 1976
prolactin levels	Serum	5 weeks	51% increased	Govoni et al., 1978

Table 9 *continued*

Neurotransmitter	Area	Age	Effect	Reference
Noradrenaline endogenous levels	Cerebellum		Increased	Krall et al., 1972
	Halfbrain	day 32	No change	Sauerhoff and Michaelson, 1973
		day 33	Increased	Michaelson et al., 1974
	Forebrain	40–90 days	Increased	Golter and Michaelson, 1975
	Whole brain		No change	Silbergeld and Goldberg, 1975
	Brainstem	day 15	No change	Schumann, 1977
	Striatum	8 weeks	Increased	Jason and Kellogg, 1977
			Increased	Dubas et al., 1978
			Decreased	Dubas and Hrdina, 1978
		100 days	No change	Rafales et al., 1981
	Cortex	8 weeks	Decreased	Dubas and Hrdina, 1978
	Midbrain	8 weeks	Increased	Dubas and Hrdina, 1978
			Increased	Dubas et al., 1978
	Hypothalamus	8 weeks	No change	Dubas and Hrdina, 1978
			Increased at 20 ppm	Dubas et al., 1978
synthesis	Whole brain		No change	Schumann, 1977
uptake	Forebrain		No change	Silbergeld and Goldberg, 1975
Adenylate cyclase activity basal activity	Striatum	5 weeks	No change	Govoni et al., 1979
	Cerebellum	day 52	18% increased	Nathanson, 1979
	Caudate nucleus	day 52	32% increased	Nathanson, 1979
after chelation	Cerebellum	day 52	48% increased	Nathanson, 1979
	Caudate nucleus	day 52	18% increased	Nathanson, 1979
	Forebrain	25–30 days	No change	Wince et al., 1980
stimulation by DA	Forebrain	25–35 days	40% decreased	Wince and Azzarro, 1978
stimulation by APO		25–30 days	Decreased	Wince et al., 1980
			Decreased	Wince et al., 1980
Spiperone specific binding	Striatum	6 weeks	No change	Lucchi et al., 1981

Table 9 *continued*

Neurotransmitter	Area	Age	Effect	Reference
phosphodiesterase	Cerebellum	day 52	12% decreased	Nathanson, 1979
Tetrahydrobiopterin endogenous levels	Plasma	3 months	Increased	McIntosh et al., 1982
		7 months	Decreased	McIntosh et al., 1982
INDOLAMINES				
5-Hydroxytryptamine endogenous levels	Brain	Adults	Increased	Weinreich et al., 1977
	Cortex	8 weeks	Decreased	Dubas and Hrdina, 1978
	Striatum	8 weeks	No change	Dubas and Hrdina, 1978
				Dubas et al., 1978
	Midbrain	8 weeks	No change	Dubas and Hrdina, 1978
			Increased at 20, 40 ppm	Dubas et al., 1978
	Hypothalamus	8 weeks	Decreased	Dubas and Hrdina, 1978
			Increased at 40 ppm	Dubas et al., 1978
	Whole brain		Increased	Gerber et al., 1978
uptake	Forebrain	40–90 days	No change	Silbergeld and Goldberg, 1975
5-Hydroxyindolacetic acid endogenous levels	Cortex	8 weeks	Decreased	Dubas and Hrdina, 1978
	Striatum	8 weeks	No change	Dubas and Hrdina, 1978
				Dubas et al., 1978
	Midbrain	8 weeks	No change	Dubas and Hrdina, 1978
			Decreased at 20, 40 ppm	Dubas et al., 1978
	Hypothalamus	8 weeks	Decreased	Dubas and Hrdina, 1978
			Decreased at 40 ppm	Dubas et al., 1978
γ-AMINOBUTYRIC ACID (GABA)				
endogenous levels	Cerebellum	30 days	Decreased	Piepho et al., 1976
		45–60 days	Reduced at 10 mg/ml	Silbergeld et al., 1979; 1980
	Brainstem	30 days	No change	Piepho et al., 1976
	Cerebrum	30 days	No change	Piepho et al., 1976

Table 9 *continued*

Neurotransmitter	Area	Age	Effect	Reference
	Cortex	45–60 days	No change	Silbergeld et al., 1979; 1980
	Caudate nucleus	45–60 days	No change	Silbergeld et al., 1979; 1980
	Substantia nigra	45–60 days	No change	Silbergeld et al., 1979; 1980
uptake	Forebrain	40–90 days	No change	Silbergeld and Goldberg, 1975; 1980
	Whole brain		Decreased	Gerber et al., 1978
	Cortex	45–60 days	Decreased	Silbergeld et al., 1979; 1980
	Caudate nucleus	45–60 days	Decreased	Silbergeld et al., 1979; 1980
	Substantia nigra	45–60 days	Decreased	Silbergeld et al., 1979; 1980
	Cerebellum	45–60 days	No change	Silbergeld et al., 1979
		45–60 days	Decreased	Silbergeld et al., 1980
			No change	Ramsay et al., 1980
accumulation	Cortex	45–60 days	Increased at 10 mg/ml	Silbergeld et al., 1979
	Caudate nucleus	45–60 days	No change	Silbergeld et al., 1979
	Substantia nigra	45–60 days	Increased	Silbergeld et al., 1979
	Cerebellum	45–60 days	No change	Silbergeld et al., 1979
release		45–60 days	Decreased	Silbergeld et al., 1979; 1980
KCl release		45–60 days	Decreased	Silbergeld et al., 1979; 1980
binding	Striatum	6 weeks	Decreased	Govoni et al., 1980
			Decreased	Memo et al., 1980b
	Cerebellum	6 weeks	Increased	Govoni et al., 1980
			Increased	Memo et al., 1980b
	Cortex	6 weeks	No change	Govoni et al., 1980
			No change	Memo et al., 1980b
	Hypothalamus	6 weeks	No change	Govoni et al., 1980
			No change	Memo et al., 1980b
	T. olfactor	6 weeks	No change	Govoni et al., 1980
	Substantia nigra	6 weeks	No change	Memo et al., 1980b
			No change	Memo et al., 1980b
	Nucleus accumbens	6 weeks	No change	Govoni et al., 1980
			No change	Memo et al., 1980b
	Pituitary	6 weeks	No change	Memo et al., 1980b

Table 9 *continued*

Neurotransmitter	Area	Age	Effect	Reference
binding sites	Striatum	6 weeks	Decreased	Memo et al., 1980b
	Cerebellum	6 weeks	Increased	Memo et al., 1980b
	Cortex	6 weeks	No change	Memo et al., 1980b
	Hypothalamus	6 weeks	No change	Memo et al., 1980b
	T. olfactor	6 weeks	No change	Memo et al., 1980b
	Substantia nigra	6 weeks	No change	Memo et al., 1980b
	Nucleus accumbens	6 weeks	No change	Memo et al., 1980b
	Pituitary	6 weeks	No change	Memo et al., 1980b
Glutamic acid decarboxylase activity	Cortex	45–60 days	No change	Silbergeld et al., 1979; 1980
	Caudate nucleus	45–60 days	Increased at 5 mg/ml	Silbergeld et al., 1979; 1980
	Substantia nigra	45–60 days	No change	Silbergeld et al., 1979; 1980
	Cerebellum	45–60 days	No change	Silbergeld et al., 1979; 1980
GABA transaminase activity	Cortex	45–60 days	Decreased at 5 mg/ml	Silbergeld et al., 1979
	Caudate nucleus	45–60 days	No change	Silbergeld et al., 1979
	Substantia nigra	45–60 days	No change	Silbergeld et al., 1979
	Cerebellum	45–60 days	No change	Silbergeld et al., 1979
Guanylate cyclase endogenous levels	Striatum	6 weeks	Decreased	Govoni et al., 1980
			Decreased	Memo et al., 1980b
	Cerebellum	6 weeks	Increased	Govoni et al., 1980
			Increased	Memo et al., 1980b
	Cortex	6 weeks	No change	Memo et al., 1980b
	Hypothalamus	6 weeks	No change	Memo et al., 1980b
	T. olfactor	6 weeks	No change	Memo et al., 1980b
	Substantia nigra	6 weeks	No change	Memo et al., 1980b
	Nucleus accumbens	6 weeks	No change	Memo et al., 1980b
	Pituitary	6 weeks	No change	Memo et al., 1980b

Table 9 *continued*

Neurotransmitter	Area	Age	Effect	Reference
NEUROPEPTIDES				
Enkephalin				
endogenous levels	Striatum	During dosing	50% increased	Govoni, 1980
	Nucleus accumbens	After dosing	No change	Govoni, 1980
	Striatum	During dosing	No change	Govoni, 1980
	Striatum	10, 21, 30 days	Decreased	Winder *et al.*, 1984b
		Development	Delayed	Winder *et al.*, 1984b

medulla-pons and midbrain in rats treated with high doses of lead. In contrast, choline acetyltransferase was significantly increased in cerebral cortex, hippocampus and medulla-pons regions. The decreased acetylcholinesterase correlated with the increased steady state acetylcholine levels in diencephalon. However, in other areas there was no correlation between enzyme activities and steady state levels of neurotransmitter. The effect of lead on the more ubiquituous enzyme plasmacholinesterase (butyrylcholinesterase) has been studied. Sobotka *et al.* (1975) reported inhibition in rats dosed by gavage at doses of lead that caused under-nutrition, though Louis-Ferdinand *et al.* (1978) reported only a transient impairment. This was concluded to be due to an indirect effect of lead on the enzyme.

Least contradictory data on the effects of lead exposure on the cholinergic system are the results of turnover studies. It would appear that lead has a consistent inhibitory effect on cholinergic metabolism. This is evident in the periphery (Silbergeld and Goldberg, 1974a; Carroll *et al.*, 1977), and in the CNS, where significant reductions were found in *in vivo* acetylcholine turnover rates in all areas investigated (Shih and Hanin, 1977; 1978a).

Most early studies of pathological change in CNS cholinergic systems ascribed to lead are difficult to interpret due to the rapid onset of post-mortem changes. Recent studies designed to alleviate this problem, by killing using microwave irradiation have shown a regional diminution of brain acetylcholine, although this effect is not dose-dependent (Modak *et al.*, 1978).

The mechanism of lead-induced changes appears to be by competitive replacement of calcium at presynaptic sites affecting acetylcholine release or the uptake of choline and thus decreasing acetylcholine release (Carroll *et al.*, 1977; Silbergeld, 1977; Silbergeld and Adler, 1977). However, the energy production inhibition reported by Holtzman and colleagues (1976; 1980; 1981) and Bull *et al.* (1979) also appears to limit availability of acetyl coenzyme A. This enzyme is essential for acetylcholine production. Sterling *et al.* (1982) reported that the incorporation of radiolabelled glucose into lactate, citrate and acetylcholine was considerably reduced in adult rats drinking 200 ppm Pb and 600 ppm Pb (producing blood leads of $31.8 \mu g$ Pb and $54.2 \mu g \cdot$ Pb/100 ml). This was concluded to be a generalized effect of lead on energy metabolism and not a specific step of glucose metabolism.

It appears that while effects on cholinergic metabolism have been reported in lead-dosed animals, the doses of lead used were high, and suitable precautions against enzymatic degradation following death were not taken. The results of Sterling *et al.* (1982), where non-specific changes due to a decrease in energy metabolism were reported, point to the cholinergic system being an unlikely target for a specific neurochemical change induced by lead.

93

Catecholaminergic effects

The effects of lead on the catecholaminergic neurotransmitter systems of the CNS have been extensively investigated, and therefore have been subdivided for ease of interpretation.

In vitro *studies*

Several aspects of the effect of lead on catecholaminergic neurochemistry, including metabolism of precursors and cofactors, synthesis, release, uptake and postsynaptic receptor activity have been studied *in vitro*.

A preliminary study of the synaptosomal conversion of tyrosine to dopamine suggested enhancement of catecholamine synthesis in the presence of lead at concentrations of 0.1 µmol/l–10 mmol/l (Wince *et al.*, 1976). However, the activity of tyrosine hydroxylase, the initial and rate limiting enzyme step in the catecholamine synthetic pathway, appeared unaffected by lead at concentrations of 1 mmol/l in both striatum and hypothalamus (Deskin *et al.*, 1980).

Inorganic lead has also been shown to block uptake and release of dopamine from synaptosomes. Concentrations of 10 µmol/l (Komulainen and Tuomisto, 1981), 50 µmol/l (Silbergeld, 1977) and 100 µmol/l (Bondy *et al.*, 1979) significantly inhibited uptake. The effect of lead on the synaptosomal release of dopamine appears less clear. According to one report only calcium-dependent release at a level of 100 µmol/l (Silbergeld, 1977) was increased; two other studies of resting release of dopamine gave conflicting results, a concentration of inorganic lead of 10 µmol/l stimulating release in one (Bondy *et al.*, 1979) and up to 100 µmol/l producing no change in another (Komulainen and Tuomisto, 1981).

Lead has been shown to block postsynaptic adenylate cyclase activity at concentrations below 3 µmol/l (Wince *et al.*, 1976). This enzyme may be coupled to dopamine receptors and alterations in its function may produce changes in dopaminergic activity.

These *in vitro* findings are hard to relate to lead effects on the intact nervous system. The most important question to be considered is the physiological significance of the concentration of lead present in the experimental system. Neonates from maternal rats consuming 4% lead carbonate in their diet (27 000 ppm Pb) and subsequently weaned to 40 ppm Pb have total brain lead concentrations in the range 2.9–4.3 µmol/l (Sauerhoff and Michaelson, 1973). While it is important to recognize that the brain is not homogeneous and that localized increases in concentrations of lead will exist in certain areas (Danscher *et al.*, 1975) or intracellular compartments (Silbergeld *et al.*, 1977), any *in vitro* studies reporting changes at concentrations above these values seem unlikely to have much relevance to *in vivo* exposure.

In vivo experiments

Early research concentrated on the effects of lead on endogenous levels of neurotransmitters. Some reports indicated increased noradrenaline in whole brain (Golter and Michaelson, 1975), forebrain (Silbergeld and Goldberg, 1975), midbrain (Dubas and Hrdina, 1978) and brainstem (Jason and Kellogg, 1977) while others indicated no changes (Sauerhoff and Michaelson, 1973; Sobotka et al., 1975; Grant et al., 1976).

In the case of dopamine, some workers found no changes in lead-exposed animals (Golter and Michaelson, 1975; Silbergeld and Goldberg, 1975; Sobotka et al., 1975; Grant et al., 1976) while others reported decreases in dopamine levels in whole brain (Sauerhoff and Michaelson, 1973) striatum (Jason and Kellogg, 1977) and cortex, midbrain and hypothalamus (Dubas and Hrdina, 1978). Many of these studies did not report lead levels. However, Schumann (1977) showed that neonates from maternal rats dosed with 5% lead acetate in the diet, with blood lead levels up to 486 µg/100 ml and brain leads up to 7.5 ppm, had no significant alterations in endogenous levels of tyrosine, dopamine or noradrenaline.

Data also exists for in vivo uptake and release of catecholamines. Silbergeld and Goldberg (1975), reported a decrease in high affinity dopamine transport and an increase in high affinity transport in mice drinking water containing 5 g Pb/l (2625 ppm Pb). However, a later study by Wince et al. (1980) on the offspring of rats fed with 4% lead carbonate weaned to 40 ppm Pb showed no differences in dopamine uptake, or tyrosine utilization. As this study used pair-fed animals to control for the effects of undernutrition, it is possible that the earlier findings were complicated by indirect effects of lead (i.e. nutritional deficiencies).

These early studies which exhibited a great degree of disparity indicated that whole brain assessment and simple estimation of monoamine concentrations are inadequate measures of change. As a result, later studies have concentrated on more specific aspects of catecholaminergic function.

Catecholamine turnover rates were reported to be unchanged except in one study where whole brain noradrenaline turnover was found to be increased (Golter and Michaelson, 1975). It should be noted that turnover studies utilize the inhibition of an enzyme of transmitter synthesis (usually tyrosine hydroxylase) and measurement of accumulation of catecholaminergic intermediates. If lead were to cause an alteration at the level of the synthesizing enzymes then these methods would be unlikely to detect any lead effect.

Another aspect of catecholaminergic turnover that has been studied is the measurement of the metabolites of dopamine (homovanillic acid, HVA) and noradrenaline (vanillylmandelic acid, VMA). Silbergeld and Chisholm (1976) reported that HVA was increased by 33% in brain and 265% in urine, and VMA was increased by 48% in brain and 216% in urine in mice given 5 g Pb/l (2625 ppm Pb) from birth. It is likely that increased urinary metabolite levels reflect peripheral catecholamine breakdown; therefore this dose of lead would appear to affect both peripheral and central metabolisms. This study also reported that the urinary concentrations of these metabolites, especially HVA, were elevated in lead-exposed children (i.e. children with blood lead in the range 59–68 μg Pb/100 ml).

In a study of dopaminergic turnover in rats dosed throughout life with 2.5 g Pb/l (1365 ppm Pb), Govoni et al. (1978) reported that dihydroxy-phenylacetic acid (DOPAC, a reliable indirect index of functional ability of dopaminergic neurones) and HVA concentrations were significantly decreased in the striatum.

In a later study using a similar dosing regime Govoni et al. (1979) reported that besides being decreased in the striatum, DOPAC activity was unchanged in the substantia nigra and increased in the nucleus accumbens and frontal cortex. The animals in both studies were reported to have increased loco-motor activity, although it should be noted that in the earlier (1978) study, lead-dosed animals were 10–15% lighter. The reported increase of DOPAC in the nucleus accumbens was concluded to be due to an enhancement of dopamine synthesis, which could be correlated with the increase in motor activity. The other changes were more difficult to interpret, and Memo et al. (1981) suggested that interactions with other neurotransmitter systems might account for them. These authors confirmed previous findings and reported differential DOPAC levels in various CNS regions. These differences were both dose dependent (at 0.04 g Pb/l and 218 ppm Pb and 2.5 g Pb/l – 1365 ppm Pb) and reversible (these changes had returned to normal 30 days after cessation of treatment).

Further reinforcement of these findings came from Lucchi et al. (1981) who reported receptor changes in striatum (increased) and nucleus accumbens (decreased) by measuring (–)[^{3}H]sulpiride stereospecific binding (a substituted benzamide believed to be a specific ligand for dopamine D2 receptors) in rats dosed throughout life with 2.5 g Pb/l (1365 ppm Pb). These findings were in the opposite direction to the alteration in DOPAC levels (Govoni et al., 1978 and 1979) and it was concluded that these changes were consistent with their results i.e. an increase in dopamine synthesis caused a decrease in receptor sensitivity and vice versa. It is pertinent to note that these authors found no differences in spiperone specific binding (a label for both D1 and D2 receptors but with preferential labelling of D2) in the areas studied. There are problems regarding the use of sulpiride as a ligand in D2 receptor binding, and these include specificity and sensitivity (Lazareno and Nahorski, 1982).

Finally, as the receptor-linked adenyl cyclase (dopamine D1) in forebrain synaptosomes (Wince et al., 1980; Wince and Azzarro, 1978); receptor-linked guanyl cyclase activity (Govoni et al., 1980); and noradrenaline-sensitive adenyl cyclase in the cerebellum (Taylor et al., 1978) have been reported to be changed, the possible receptor alterations induced by lead remain inconclusive.

Little work has been carried out on noradrenaline, the other major central catecholamine neurotransmitter. Goldman *et al.* (1980) dosed maternal rats with drinking water concentrations of 0.05% (278 ppm Pb), 0.1% (545 ppm Pb) and 0.2% (1090 ppm Pb) from birth till weaning. These gave blood leads at 21 days of 13 μg, 21 μg and 41 μg Pb/100 ml respectively. Increases in serum noradrenaline levels (220%) and adrenal noradrenaline levels (53%) were found. Elevations in brainstem dopamine β-hydroxylase (the enzyme that converts dopamine to noradrenaline) and phenylethanolamine-N-methyl transferase (the enzyme responsible for the methylation of noradrenaline to adrenaline) of 36% and 46% were observed. These findings support the evidence for an increased central catecholaminergic turnover at low levels of lead exposure, although a possible contribution from peripheral catecholaminergic metabolism cannot be ruled out.

In summary, two types of lead effect have been reported on catecholaminergic neurochemistry. The first of these is shown mainly in the earlier studies, where relatively large exposures of lead were employed, causing a variety of changes. In these studies, catecholaminergic function was found generally to be inhibited, perhaps as a result of non-specific effects of lead, including general toxicity and undernutrition.

The second type of effect has been seen in studies utilizing relatively small doses of lead and monitoring lead load by measuring blood lead levels. The results of these studies are harder to interpret, but overall they suggest a stimulation of catecholaminergic function at levels of lead comparable to those found in exposed humans. The results of low-dose experiments in which drug-elicited behaviour is studied are of particular importance in the context of such borderline effects.

Effects on γ-amino butyric acid (GABA)

Early studies of lead on GABA metabolism measured simple parameters such as endogenous levels in crude brain areas using high levels of lead. No change was found in whole forebrain GABA levels of rats or mice by Sauerhoff and Michaelson (1973), although Piepho *et al.* (1976) reported a decrease in GABA levels in cerebellum (but not cerebrum or brain stem) of 30-day-old rats dosed by gavage with 27 mg and 54 mg Pb/kg, producing blood leads of 56.2 μg and 128 μg Pb/100 ml.

Silbergeld *et al.* (1979) exposed rats to drinking water containing 5 mg or 10 mg Pb/ml (2625 ppm or 5450 ppm Pb) from birth. Nevertheless, exposure sensitized them to the behavioural effects of convulsant agents (see below), and several aspects of GABAergic function. These included inhibition of normal and potassium-stimulated GABA uptake and release in all areas but the cerebellum; decrease in GABA levels in the cerebellum; changes in GABAergic enzymes such as GABAtransaminase (decreased in cortex) and glutamic acid decarboxlase (increased in substantia nigra); and increase in the apparent rate of GABA synthesis.

This work was extended (Silbergeld *et al.*, 1980) in an attempt to compare *in vivo* effects with *in vitro* effects. Sensitivity to convulsant drugs was observed again, as were increases in GABA levels in cerebellum and glutamic acid decarboxylase in caudate nucleus. However, increases in *in vitro* lead concentration up to 0.1 μmol/l had no effect on glutamic acid decarboxylase activity in brain homogenates. GABA uptake into synaptosomes from *in vivo* exposed animals was inhibited in all areas measured except cortex, but *in vitro* concentrations up to 0.1 μmol/l had no effect. Effects on resting and potassium-stimulated release of GABA from synaptosomes were also seen *in vivo* but not *in vitro*. The lack of correlation of *in vitro* to *in vivo* function was suggested as indicating that *in vivo* effects were secondary to neurotoxicity or even non-neural actions.

Govoni *et al.* (1980) found no differences in the activity of the metabolic enzymes glutamic acid decarboxylase and GABAtransaminase. This contrasts with the findings of Silbergeld *et al.* (1979) and may reflect the lower doses of lead used. Govoni *et al.* (1980) found that radiolabelled GABA binding (was increased in the cerebellum and decreased in the striatum and similar changes were also noted in cyclic GMP levels.

GABA is a major inhibitory neurotransmitter. Explanations of the consequences of these changes are therefore complicated by possible interactions with excitatory systems. An effect of lead on GABAergic neurotransmission may therefore be interpreted in two ways. Lead might generally increase excitability, which might confer sensitivity to any drug administration that reduces inhibition. Alternatively, lead might selectively interfere with specific neurochemical pathways that mediate inhibition in the CNS.

There is also the possibility that the disruption in haem biosynthesis induced by lead with a concomitant increase of D-amino laevulinic acid (ALA) may produce direct neurotoxic effects. Silbergeld and Lamon (1980) have suggested that the neurotoxicity of lead might be due in part to competitive interactions involving ALA at GABA receptors. Not only is ALA a weak displacer of specific GABA binding to synaptic membranes (Silbergeld and Lamon, 1980), but also it directly inhibits GABA release from synaptosomes (Brennan and Cantrill, 1979).

Effects on other neurotransmitters

The effects of lead on the opioid neurotransmitters, the enkephalins, has recently been studied. An increase in striatal enkephalins has been observed following a three month period of continuous lead exposure (Govoni *et al.*, 1980; Memo *et al.*, 1980c). Studies using lower and shorter lead exposures have shown a depression of enkephalin levels in the striatum and a delay in the ontogeny of this neurotransmitter (Winder *et al.*, 1984b).

Dubas and Hrdina (1978) and Dubas *et al.* (1978) reported that endogenous levels of 5-hydroxytryptamine were decreased in cortex and hypothalamus, and levels of its metabolite 5-hydroxyindole acetic acid were decreased in cortex, midbrain and hypothalamus.

Levels of various amino acids, including some that are considered putative neurotransmitters have also been measured in lead-dosed animals. The normal developmental increase of glutamate, aspartic acid and glutamine were found to be decreased in animals dosed with the Pentschew and Garro (1966) model (Patel *et al.*, 1974a). Forebrain uptake of tyrosine (Silbergeld and Goldberg, 1975) and whole brain uptake of proline (Gerber *et al.*, 1978) have been found to be increased. Uptake in forebrain of other amino acids (glycine, phenylalanine and leucine) were unchanged (Silbergeld and Goldberg, 1975). The functional implications of these changes are not clear, especially as some of the doses of lead used were high.

The possibility that variations in catecholamine concentrations are compensatory responses to changes in postjunctional processes (i.e. the adenylate cyclase-linked dopamine receptor) cannot be excluded, as it has been shown *in vitro* that adenylate cyclase activity and cyclic AMP levels are decreased by lead (Nathanson and Bloom, 1975). Adenylate cyclase activity was also depressed *in vivo* in acute (high dose) experiments although it was not significantly changed in animals which had been chronically dosed (Ewers and Erbe, 1980). The alteration in GABA and enkephalin systems discussed above may also cascade onto other neurotransmitter systems.

SUMMARY

It is not unexpected that heavy metal ions such as lead produce toxicity. Metabolic machinery is not present to deal with their presence and though the absorption of heavy metals is relatively poor, they persist in the body and produce cumulative toxicity. A great deal of the studies cited in this chapter have used lead dosing models which produce high blood lead levels and it is not surprising that disorders of a biochemical, neurochemical or behavioural nature have been observed. What is of more interest and concern are those studies that have used lower lead dosing regimes which give blood lead levels in animals that fall within the limits observed in humans with subclinical lead intoxication. These studies may give some insight into the behavioural disorders in children (such as hyperactivity or mental retardation), which may have been correlated with environmental exposure to lead.

Of the biochemical abnormalities seen in lead-treated animals, the most important is the alteration in oxidative metabolism. These have been shown at high levels, but more relevantly, at lower levels of lead exposure in both adult rats (Sterling *et al.*, 1982) and neonates (Bull *et al.*, 1979). These have obvious repercussions for other aspects of CNS function. Also, biopterin metabolism appears to be one of the biochemical systems most sensitive to lead, comparable to D-amino laevulinic acid synthesis, although the details of the system have not been fully investigated. This enzyme system produces cofactors for aminergic synthesis (including the catecholaminergic and 5-hydroxytrytaminergic systems) and disruption may cause serious neurochemical consequences.

Neurochemically, the most investigated transmitter has been dopamine, where there may be some specific changes at low exposure levels. The reported changes in the development of the enkephalin system also appear to occur at low levels. Other systems, such as the cholinergic or GABAergic, appear to show what are relatively high dose non-specific changes in metabolism or function. Other possible or putative neurotransmitters have not been sufficiently investigated and it is not possible to make firm conclusions regarding the possible effects of these.

The effects of lead on drug-induced behaviour may also give information concerning neurochemical function. As noted in Table 7d and Chapter 4, the drug most frequently administered to lead-dosed animals has been amphetamine. The responses to this drug appear to be affected by the exposure of lead given. At high levels, amphetamine-induced behaviour appears to be paradoxical or attenuated, but at lower lead doses it is possible that the dopamine system is stimulated by this drug. Of other drug-stimulated behaviour, sensitivity of response appears to be enhanced by lead in the GABAergic system.

Lastly, some attempts have been made to correlate behavioural findings with those reported at the neurochemical level. These have not been wholly successful (e.g. Memo et al., 1980) but some links have been reported (e.g. Silbergeld et al., 1979; Jason and Kellogg, 1981). This is worthy of further attention.

6
The Morphological Effects of Lead

THE CENTRAL NERVOUS SYSTEM

The neuropathology of lead intoxication cannot readily be studied in humans, except in fatal cases with lead encephalopathy, where tissue becomes available at autopsy. In such cases, oedema and haemorrhage, with secondary glial and neuronal changes may be found. It is apparent that striking neuropathological alterations of a primarily nature are seen in the brains of human subjects exposed to high levels of lead. Clinical observations suggest the possibility of specific nerve cell lesions at high lead levels, but pathological identification of such lesions is lacking. The neuropathological changes, if any, that may be present in the central nervous system of humans chronically exposed to low lead levels remain to be defined. Overall, knowledge of the neuropathology of lead intoxication remains incomplete, and it could still be stated, in the words of Kinnier Wilson, that 'from a neuropathological aspect many of the older records in a scattered literature are too meagre for present-day standards, while those providing needful information are sometimes contradictory. The indefiniteness which still surrounds the subject is rather remarkable' (Wilson, 1954). Because of the rarity today of lead encephalopathy, the older pathological studies are unlikely to be superceded. Although undoubtedly widespread, low level lead intoxication in man is also unlikely to be a source of usable new pathological material. However imperfect, the historical reports summarized by Wilson (1954), Pentschew (1965), Hirano and Iwata (1979) and others would therefore appear to be definitive. Increasingly, the understanding of the effects of lead on the human nervous system will depend on findings in experimental animals.

PERIPHERAL NERVOUS SYSTEM

In man, damage to peripheral nerves in occupationally exposed individuals has been well studied from both clinical and pathological standpoints. Involvement of the muscles of the forearm, and to a lesser degree the upper arm, had as its pathological counterpart changes in peripheral nerves, evidence of muscle atrophy of the type seen with nerve damage, and degenerative–reactive appearances of nerve cells in spinal cord. These appearances are all indicative of a primary action of lead on neurones, rather than on the paralysed muscles themselves. While high level occupational lead exposure undoubtedly causes neuronal damage, firm evidence for such damage at lower levels is unavailable. On the basis of measurements of nerve impulse conduction it has been claimed (by Seppalainen *et al.*, 1975) that lead workers with blood levels below 70 μg/100 ml have signs of nerve damage. However, this finding has not proved reproducible, and both techniques and interpretation have been strongly criticized. Other studies of nerve conduction, both in adults and lead-exposed children, are bedevilled by problems of wide variability of conduction velocity in normal individuals (see Cooper and Sigwart, 1980), so that, on currently available data, it is not yet possible to conclude that low-level lead exposure damages the nerves in the limbs.

Lead-induced peripheral neuropathy has been extensively studied in laboratory animals. In adults the limb nerves appear more readily affected than the brain at high exposure levels, so that there are ample experimental findings to complement clinical observations on lead paralysis. The most prominent change in limb nerves is damage to myelin sheaths, a phenomenon first observed by Gombault (1880). Loss of myelin is often visibly segmental, implying damage in the supportive Schwann cells. Using a high-dose, prolonged exposure regime, Lampert and Schochet (1968) confirmed that in guinea pigs the most obvious effects of lead were on these cells. Degeneration of Schwann cells was associated with myelin breakdown, while evidence of their regeneration was found, together with signs of remyelination, suggesting intermittent recovery. The possibility that Schwann cell and myelin damage might occur secondarily to damage to blood vessel walls was also raised. They noted that the nerves of lead-exposed animals were swollen, apparently due to leakage of fluid from blood vessels with visibly abnormal lining cells. Dyck *et al.* (1980) further examined the question of a 'blood–nerve barrier' breakdown in rats fed 4% lead carbonate for up to 12 weeks. Although nerve swelling was seen, the passage of two protein marker substances through blood vessel walls was increased either minimally or not at all. Their observations did not add weight to the hypothesis that myelin damage in nerves of lead-treated animals is at least in part dependent on massive leakage of lead through altered blood vessels, though the passage of proteins across vessel walls does not necessarily mirror that of small molecules like lead.

Lead-exposed nerves have often been reported to show evidence of damage in their axonal fibres as well as in myelin sheaths. Schlaepfer (1969) has shown this phenomenon. He exposed adult rats massively to lead (2.4–3.0 g lead acetate/kg/day for between 3 and 18 months). Characteristic myelin sheath changes were noted in hindlimb nerves, while Wallerian degeneration, indicative of axonal breakdown, was seen in some animals. The supportive cells in the anterior horn of the spinal cord were completely normal, a finding confirmed quantitatively by Ohnishi *et al.* (1977); however, reactive proliferation of capsule cells around sensory neurones in dorsal root ganglia was observed. Axonal breakdown is evidently a minor element in lead-induced nerve damage, but it has been described consistently enough for a lead effect on peripheral nerve axons and their parent nerve cells to be adduced.

In young animals effects of lead may be seen on peripheral nerves but are overshadowed by effects on the central nervous system. Brashear *et al.* (1978) exposed suckling rats to 1 g lead (as acetate)/kg/day by gavage from days 2 to 20. Survivors after 15 days showed residual haemorrhagic damage in brain and spinal cord. Segmental demyelination was not produced in peripheral nerves, though myelin formation showed ultrastructural abnormalities. It is not certain that the severe undernutrition induced by the high lead dose might not have been responsible for this effect. Toews *et al.* (1980) also caused undernutrition in rats with a high level of lead, and described reduced brain myelination. In contrast, sciatic nerve myelin accumulation was not affected. It appears that in developing animals the relative susceptibilities of peripheral nerves and central nervous system are the opposite of what is found in adults.

In summary, lead affects peripheral nerves by damaging Schwann cells and myelin sheaths, by damaging some axons (though not definitely their precursor cells), and by influencing the permeability of blood vessels within nerves. The primary site of damage is not known, nor is it known what effects, if any, may be seen at levels of exposure comparable to those found in the contemporary urban environment.

NEUROPATHOLOGY OF LEAD IN ANIMALS

The relative susceptibility to lead of the developing brain was a fact exploited successfully by Pentschew and Garro (1966). Their model system for lead encephalopathy used a maternal diet containing 4% lead carbonate fed to rats from parturition. As lead crosses into the milk of a lactating animal, a developing neonate will receive lead from this source. Maternal rats were apparently unaffected, but the 4.59 mg/100 ml (45.9 ppm) of lead in their milk was sufficient to produce pronounced retardation in growth of the offspring. Towards the end of the suckling period, severe abnormal physical signs developed (hind limb paralysis, urinary incontinence, ruffled coat) in 90% of the neonates with subsequent death in nearly all cases. Those animals which survived the first 2 weeks eventually recovered, even when maintained on the lead-supplemented diet.

Macroscopically, the cerebellum was the area of the brain most severely affected, although damage was seen in the cerebral hemispheres. A reddish brown discolouration was the most obvious change in the cerebellum, and closer examination revealed petechial haemorrhage and oedema. The fore-brain was pale with occasional haemorrhages. Microscopic examination showed prominent capillaries, endothelial and glial proliferation, and spongy changes in the neuropil. Capillary permeability was also altered: trypan blue, injected systemically into 20-day-old animals, was found to penetrate cerebral vessels diffusely.

Lampert et al. (1967) re-examined these findings under the electron micro-scope. They found that the blood–brain barrier was functionally absent, and that many capillary endothelial cells were vacuolated. In addition, the extracellular spaces in both grey and white matter were abnormally wide, although this disturbance was only transient in grey matter. Glial cells had proliferated, with an increase in astrocyte filaments as well as swelling and other reactive changes.

Since these pioneering studies, a great deal of work has been carried out and a substantial body of literature has accumulated on the morphological and structural effects of lead. Large doses of lead administered to developing rats reproducibly cause vascular changes, and it is not necessary for lead to be administered strictly in accordance with the Pentschew–Garro model. Alternative methods of administration and variable doses produce com-parable effects. In recent years, interest in the clinical problems related to low-level lead exposure has to some extent directed experimental attention away from high-dose systems with vascular encephalopathic changes to others where lower doses of lead are given in an attempt to identify less dramatic neurotoxic effects.

PATHOLOGICAL FINDINGS IN LEAD-DOSED ANIMALS

High versus low dose studies

The published literature on lead-induced neuropathological changes in experimental animals, primarily rats, is now extensive. The papers reviewed in Table 10 have been examined critically where possible, with the exclusion of some alleged findings, and association of changes with lead treatment should not be taken to imply a causal relationship. The changes observed have been divided into those found in high and low dose studies.

High doses produce vascular encephalopathic changes, with effects on neurones and glia which are reactive to the consequences of blood vessel damage (but probably not exclusively so). With high and intermediate dose levels, the effects of lead on the CNS will be modulated by concomitant somatic changes induced as a consequence of significant nutritional deficits (Michaelson and Sauerhoff, 1974b). Undernutrition itself is a powerful cause of neuropathological change (c.f. Balazs et al., 1979) and it may not be possible to separate such an indirect effect from more specific effects of lead. Therefore high dose effects have been defined not only as changes related to

Table 10 Morphological effects of developmental lead exposure in rats and mice

Parameter	Effect	High dose studies (with undernutrition)	Effect	Low dose studies (no undernutrition)
WHOLE BRAIN				
Weight	Decreased	Krigman et al., 1972; 1974a; b		
	Decreased	Krigman and Hogan, 1974		
	Decreased	Press, 1977a		
	Decreased	Toews et al., 1978		
	Decreased	Petit and Alfano, 1979		
	Decreased	Petit and LeBoutillier, 1979		
	No change	Barlow et al., 1977		
	Decreased	Carmichael et al., 1981	No change	Carmichael et al., 1981
	No change	Petit and Alfano, 1979	No change	Campbell et al., 1982
	No change	Miller et al., 1982	No change	Miller et al., 1982
	No change	Miller et al., 1982	No change	Miller et al., 1982
Total DNA (cell number)				
FOREBRAIN				
Weight	Decreased	Michaelson and Sauerhoff, 1974		
	Decreased	Petit and Alfano, 1979		
	Decreased	Toews et al., 1980		
	No change	Petit and Alfano, 1979		
	No change	Michaelson and Sauerhoff, 1974		
Total DNA (cell number)				
CEREBRUM				
Cortical development	No change	Krigman et al., 1974b		
Cortical thickness	Decreased	Krigman et al., 1972; 1974b		
	Decreased	Reyners et al., 1976		
	Decreased	Petit and LeBoutillier, 1979		
Vascular changes				
oedema	Present	Michaelson and Sauerhoff, 1974		
	Present	Ahrens and Vistica, 1977		
haemorrhage	Present	Ahrens and Vistica, 1977		
	Present	Minsker et al., 1979		
	Present	Toews et al., 1978		

Table 10 *continued*

Parameter	Effect	High dose studies (with undernutrition)	Effect	Low dose studies (no undernutrition)
capillary density	Present	Alfano et al., 1982		
	No change	Reyners et al., 1979		
capillary permeability	Increased	Reyners et al., 1976		
	Increased	Pentschew and Garro, 1966		
	Increased	Vistica and Ahrens, 1977		
	Increased	Petit and Alfano, 1979		
endothelial swelling	Present	Ahrens and Vistica, 1977		
intercellular junctional changes	Present	Tennekoon et al., 1979		
basement membrane development	Retarded	Vistica and Ahrens, 1977		
		Vistica and Ahrens, 1977		
Glial changes				
total density	Increased	Krigman et al., 1974b		
	Increased	Krigman and Hogan, 1972; 1974		
astrocytes	No change	Reyners et al., 1979		
	Increased	Krigman et al., 1974b		
	Increased	Press, 1977b		
oligodendrocytes	Decreased	Reyners et al., 1979		
microglia	No change	Reyners et al., 1979		
	Increased	Reyners et al., 1978		
Nerve cell changes				
packing density	Increased	Krigman et al., 1972; 1974b		
	Increased	Krigman and Hogan, 1974		
axon size	Decreased	Krigman et al., 1974b		
diameter of processes	Increased	Krigman et al., 1974b		
	Increased	Krigman and Hogan, 1974		
nuclear size	Decreased	Krigman et al., 1972; 1974b		
synapses/neuron	Decreased	Krigman et al., 1974b		
	Decreased	Krigman et al., 1980		
synaptic density	No change	Krigman et al., 1974b		
	No change	Krigman and Hogan, 1974		

Table 10 *continued*

Parameter	Effect	High dose studies (with undernutrition)	Effect	Low dose studies (no undernutrition)
presynaptic dense projections				
size	Decreased	Petit and LeBoutillier, 1979	Decreased	McCauley et al., 1979
	Decreased	Krigman et al., 1980		
	Decreased	Alfano and Petit, 1982		
boutons	No change	Petit and LeBoutillier, 1979	Decreased	McCauley et al., 1979
	Decreased	Krigman et al., 1972		
	Increased	Krigman et al., 1974b		
synaptic cleft width	No change	Petit and LeBoutillier, 1979		
postsynaptic dimensions	No change	Petit and LeBoutillier, 1979		
dendritic development	Decreased	Petit and LeBoutillier, 1979		
dendritic arborization	Decreased	Petit and LeBoutillier, 1979		
HIPPOCAMPUS				
Histological appearances	No change	Minsker et al., 1979	No change	Minsker et al., 1979
Pyramidal cell				
layer thickness	No change	Louis-Ferdinand et al., 1978	No change	Louis-Ferdinand et al., 1978
	No change	Minsker et al., 1979; 1982	No change	Minsker et al., 1979; 1982
spine density	Decreased	Kiraly and Jones, 1982		
Mossy fibre				
pathway development	Retarded	Alfano et al., 1982		
synaptic profiles			Decreased	Campbell et al., 1982
Granule cell				
layer thickness	No change	Minsker et al., 1979; 1981	No change	Minsker et al., 1979; 1982
			Decreased	Louis-Ferdinand et al., 1978
dendritic development	Decreased	Alfano and Petit, 1982	Decreased	Campbell et al., 1982
dentate hilus area			No change	Minsker et al., 1982
Hippocampal–neocortical distance			Decreased	Louis-Ferdinand, 1978

Table 10 *continued*

Parameter	Effect	High dose studies (with undernutrition)	Effect	Low dose studies (no undernutrition)
CORPUS STRIATUM				
Oedema	Present	Pentschew and Garro, 1966		
	Present	Lampert *et al.*, 1967		
Haemorrhage	Present	Pentschew and Garro, 1966		
	Present	Lampert *et al.*, 1967		
	Present	Petit and Alfano, 1979		
Capillary proliferation	Present	Pentschew and Garro, 1966		
Glial proliferation	Present	Pentschew and Garro, 1966		
CORPUS CALLOSUM				
Cyst formation	Present	Krigman and Hogan, 1974		
Oedema	Present	Holtzman *et al.*, 1980		
Haemorrhage	Present	Holtzman *et al.*, 1980		
OPTIC NERVE				
oligodendrocytes	No change	Tennekoon *et al.*, 1979		
myelination	Decreased	Tennekoon *et al.*, 1979		
axon size	Decreased	Tennekoon *et al.*, 1979		
RETINA				
nerve cell changes			Present	Santos-Anderson *et al.*, 1980
Mueller cell changes			Present	Santos-Anderson *et al.*, 1980
HINDBRAIN				
CEREBELLUM				
Weight	No change	McConnell and Berry, 1979		
	Decreased	Krigman and Hogan, 1974		
	Decreased	Michaelson and Sauerhoff, 1974		
Total DNA (cell number)	Decreased	Michaelson and Sauerhoff, 1974		
Brown discolouration	Present	Pentschew and Garro, 1966		

108

Table 10 *continued*

Parameter	Effect	High dose studies (with undernutrition)	Effect	Low dose studies (no undernutrition)
	Present	Lampert *et al.*, 1967		
	Present	Thomas *et al.*, 1971		
	Present	Krigman *et al.*, 1974a; b		
	Present	Krigman and Hogan, 1974		
	Present	Michaelson and Sauerhoff, 1974		
	Present	Press, 1977b		
	Present	Toews *et al.*, 1978		
	Present	McConnell and Berry, 1979		
Vascular changes				
oedema	Present	Pentschew and Garro, 1966		
	Present	Thomas *et al.*, 1971		
	Present	Clasen *et al.*, 1974		
	Present	Goldstein *et al.*, 1974		
	Present	Michaelson and Sauerhoff, 1974		
	Present	Thomas and Thomas, 1974		
	Present	Ahrens and Vistica, 1977		
	Present	Press, 1977a; b; c		
	Present	Toews *et al.*, 1978		
	Present	McConnell and Berry, 1979		
	Present	Holtzman *et al.*, 1980		
cyst formation	Present	Pentschew and Garro, 1966		
	Present	Krigman *et al.*, 1974a		
	Present	Krigman and Hogan, 1974		
	Present	Reyners *et al.*, 1976		
	Present	Press, 1977b		
haemorrhage	Present	Pentschew and Garro, 1966		
	Present	Lampert *et al.*, 1967		
	Present	Thomas *et al.*, 1971		
	Present	Goldstein *et al.*, 1974		
	Present	Krigman *et al.*, 1974a		

Table 10 *continued*

Parameter	Effect	High dose studies (with undernutrition)	Effect	Low dose studies (no undernutrition)
	Present	Krigman and Hogan, 1974		
	Present	Thomas and Thomas, 1974		
	Present	Ahrens and Vistica, 1977		
	Present	Press, 1977a; b; c		
	Present	Toews et al., 1978		
	Present	Holtzman et al., 1980		
	Present	LeFauconnier et al., 1980		
capillary proliferation	Present	Goldstein et al., 1974		
capillary thrombi	Present	Krigman and Hogan, 1974		
	Present	Thomas et al., 1971		
capillary necrosis	Present	Thomas and Thomas, 1974		
basement membrane changes	Present	Thomas et al., 1971		
	Present	Press, 1977b		
	Present	Vistica and Ahrens, 1977		
	Present	Holtzman et al., 1980		
Glial changes				
astrocytes	Increased	Pentschew and Garro, 1966		
	Increased	Lampert et al., 1967		
	Increased	Thomas et al., 1971		
	Increased	Krigman et al., 1974a		
	Increased	Krigman and Hogan, 1974		
	No change	Holtzman et al., 1980		
oligodendrocytes	Present	Lampert et al., 1967		
microglia	Present	Pentschew and Garro, 1966		
	Present	Lampert et al., 1967		
	Present	Thomas et al., 1971		
	Present	Goldstein et al., 1974		
	Present	Krigman et al., 1974a		
	Present	Krigman and Hogan, 1974		
	Present	Thomas and Thomas, 1974		

Table 10 *continued*

Parameter	Effect	High dose studies (with undernutrition)	Effect	Low dose studies (no undernutrition)
Neuronal changes				
Purkinje cell changes	Present	Press, 1977b		
	Present	Holtzman et al., 1980		
	Present	Pentschew and Garro, 1966		
	Present	Lampert et al., 1967		
	Present	Thomas et al., 1971		
	Present	Thomas and Thomas, 1974		
	Present	Press, 1977c		
Purkinje cell number	No change	McConnell and Berry, 1979		
Purkinje cell density	No change	McConnell and Berry, 1979		
	Decreased	McConnell and Berry, 1979		
Purkinje cell size	Increased	McConnell and Berry, 1979		
	Decreased	Thomas et al., 1971		
Purkinje dendritic topology	Abnormal	McConnell and Berry, 1979		
total segment number	Decreased	McConnell and Berry, 1979		
distal segment length	Decreased	McConnell and Berry, 1979		
percentage trichotomy	No change	McConnell and Berry, 1979		
spine network	No change	McConnell and Berry, 1979		
dendritic density	No change	Press, 1977c		
synaptogenesis	No change	McConnell and Berry, 1979		
	Decreased	McConnell and Berry, 1979		
	No change	Press, 1977c		
granule cell density	No change	McConnell and Berry, 1979		
	Decreased	McConnell and Berry, 1979		
granule cell number	No change	McConnell and Berry, 1979		
granule cell degeneration	Present	Miller et al., 1982		
molecular layer thickness	Decreased	Press, 1977c		
external granular layer cell death	Increased	Press, 1977c		

Table 10 *continued*

Parameter	Effect	High dose studies (with undernutrition)	Effect	Low dose studies (no undernutrition)
cell proliferation	Increased	Holtzman *et al.*, 1980		
foliar development	Decreased	Press, 1977c		
choroid plexus	No change	Press, 1977b		
	No change	Press, 1977b		
BRAINSTEM				
Pyramidal tract				
haemorrhage	Present	Press, 1977a		
axon size	Decreased	Krigman *et al.*, 1974a		
	Decreased	Krigman and Hogan, 1974		
myelin lamellae	Decreased	Krigman *et al.*, 1974a		
	Decreased	Krigman and Hogan, 1974		

vascular damage of the encephalopathic type; but also as those seen in animals significantly undernourished, as evidenced by reduced weight gain during the neonatal period.

Low dose studies are defined as those which employ lead levels which do not induce significant undernutrition. With the shift in research emphasis to evaluation of problems associated with low level lead in an attempt has been made to identify changes in low dose systems where non-specific influences are likely to be small, but where effects on CNS development may nevertheless occur.

HIGH DOSE STUDIES

As can be seen from Table 10 the number of studies employing high doses of lead is very large. To minimize confusion, the discussion of their reported findings has been divided into three sections, dealing with vascular, glial and neuronal effects. High doses of lead tend to cause weight loss in lead-dosed animals, and whenever reported this has been identified.

Vascular effects

Other studies in the next few years confirmed and enlarged upon the original work of Pentschew and colleagues. Thomas *et al.* (1971) suggested that one of the earliest insults inflicted by the lead treatment was to the cerebellar capillary endothelial cells. Observed effects included swelling of the cell body, vacuolation of mitochondria and endoplasmic reticulum, and increase in pinocytotic vesicles. They suggested that these effects could be the cause of the other changes observed, including deposition of degenerating endothelial cells within capillaries, collapsed vessels, 'lakes' of oedema fluid and shrinkage of neurones. A later study (Thomas *et al.*, 1973), using radiolabelled lead showed lead to be deposited within endothelial cell cytoplasm. This work has been supported by later reports (Stumpf *et al.*, 1980). Further studies (Goldstein *et al.*, 1974) found that swelling occurred in cerebral capillary endothelial cells without associated oedema or neuronal necrosis, supporting the view that (at high lead doses) the capillary endothelium was the primary site of damage.

Clasen *et al.* (1974), found developmental changes in endothelial cells (causing 'vascular strands'), and reported the presence of PAS-positive globules in perivascular astrocytes of the cerebellum and basal ganglia. These droplets, originally described by Blackman in 1936, were described as small accumulations of phagocytosed oedema fluid, possibly containing lead. The oedema fluid in the white matter was found to contain albumen and sodium. These 'vascular strands' were noted in both rat and human material, and were believed to be collapsed capillaries due to arrested development, although Reyners *et al.* (1976) considered them to be artefactual (tangential sections of capillaries).

Goldstein *et al.* (1974) described the oedematous state of abnormally increased water and sodium content in the entire brain as the prehaemorrhagic stage of lead encephalopathy.

The method of directly feeding neonate rats via an oesophagal tube (gavage) was used by Press (1977a). He found petechial haemorrhages in the cerebellar molecular layer three days after initiation of dosing. These increased in size and number to produce a severely haemorrhagic cerebellum, and similar changes appeared in the initial stages in the cerebrum and brain stem by the fifth day. By day 10, the brain was reduced in size, and the cerebellum contained large fluid-filled cavities. Cerebellar endothelial cells showed swelling and some necrosis, and Golgi preparations revealed a defective growth of capillary buds. Press (1977b) concluded that the earliest (i.e. primary) effect of lead (at high doses) was on capillary development.

Ahrens and Vistica (1977) reported retardation of neonate growth and microscopic evidence of haemorrhage and oedema, although hind limb paralysis was not observed at concentrations as high as 4% lead carbonate in diet, or 2% lead acetate in drinking water. In a companion paper on the ultrastructural effects of lead in these animals, Vistica and Ahrens (1977) supported the previous reports of vascular changes, and noted a dose-dependent delay in the developing capillary basement membranes, thereby rendering them more susceptible to the toxic effects of lead.

In a longterm dosing experiment (1.8% lead acetate in the diet, from 8 days before parturition till 3 months of age), Reyners *et al.* (1976), showed that the capillaries of the cerebral cortex of dosed neonate rats significantly increased, while the thickness of the cortex reduced progressively. In a later paper these authors reported that the absolute number of cortical capillaries was not modified by lead (Reyners *et al.*, 1979). They concluded that quantitative changes in the vascular supply represented a sequel to a primary effect on grey matter. This is in conflict with the conventional interpretation, i.e. that the neuronal changes observed are the result of a vascular upheaval.

Toews *et al.* (1978) using gavage, also found haemorrhage and oedema, with marked distortion in vascular morphology. In addition, lead was highly concentrated in isolated capillaries, even when the total level in the brain was low. When exposure was continued for 20 days, the rats recovered, despite continued high levels of lead in blood and capillaries. This suggested that adaptation to lead occurred once vascular integrity was established.

The retardation in body growth consistently observed by many workers is obviously an important lead effect. However, it is difficult to separate the effects of undernutrition (produced by generalized somatic toxicity and by anorexia) from those biochemically induced by lead. By using acute intraperitoneal injections of lead, LeFauconnier *et al.* (1980) demonstrated that haemorrhages in the cerebellum and striatum could be macroscopically observed before any weight loss. These authors also reported that animals sustained on the Pentschew and Garro diet, with 50% weight loss at 21 days, had not developed cerebellar haemorrhages. There was also a good correlation between brain lead and intensity of haemorrage. The implication of this work is that lead has a direct effect on CNS capillaries once it reaches a sufficiently high concentration, independent of any general toxicity.

Glial cell effects

The reported microscopic features of lead encephalopathy include glial proliferation in various regions of the brain. Pentschew and Garro (1966) reported intense glial proliferation in the striatum and in the foliar white matter and molecular layer of the cerebellum. When the animals had recovered, after withdrawal of lead from their diets for 2–3 weeks, gliosis disappeared. This was confirmed by other workers (Lampert et al., 1967; Krigman et al., 1972; 1974a; Krigman and Hogan, 1974). Astrocytes, especially those in the white matter, contained an increased number of filaments, mitochondria and ribosomes in their cytoplasm, and foot processes were occasionally vacuolated or entirely absent. The cytoplasm in microglia contained much phagocytosed matter, including breakdown products of red blood cells. Oligodendrocytes appeared normal in number and morphology (Lampert et al., 1967).

Replacement of satellitic oligodendrocytes by microglia was described by Reyners et al., (1978), while in a later paper, doses of lead above 5000 ppm were said to produce an increase in density of microglia and astrocytes, accompanying a reduction of oligodendrocytes (Reyners et al., 1979).

A reduction in total myelin and protein content is a further feature of acute lead encephalopathy (Krigman et al., 1972; 1974a; Krigman and Hogan, 1974). Biochemical analysis revealed that the myelin present in the lead-dosed rats was normal in composition (Krigman et al., 1974a). Except for some initiation of lamellae on unusually small axons (Krigman and Hogan, 1974), myelination in the CNS and optic nerve appeared normal, and degenerate myelin was not observed (Tennekoon et al., 1979).

The observed degree of hypomyelination appeared upon thorough morphometric analysis to be consistent with the retardation of neuronal growth and development, (Krigman and Hogan, 1974). Tennekoon et al., (1979), finding hypomyelination in the optic nerves of lead-injected mice, supported the view that this was a function of delayed neuronal development.

However, Toews et al. (1980), using rats receiving lead via gavage (100 μg or 400 μg Pb/g body weight/day), showed that accumulation of myelin in forebrain of high dosed animals was severely reduced (42%), with only a 21% reduction in brain weight. Accumulation of myelin was also reduced by 30%

in the optic nerve but unaffected in the sciatic nerve. The deficits were too great to be accounted for by undernutrition (as compared with under-nourished controls); and were not due to a developmental delay (as assessed by mature protein composition).

The basis for the effect of high levels of lead on myelination is still not clear. Whether the reported effects of lead upon myelination are due to neuronal influences, actions on the myelinating elements themselves, or as the compounded result of lead and undernutrition, remains to be resolved.

Neuronal effects

Neuronal effects of lead have been well documented. Krigman *et al.* (1974b) reported decrease in the complexity of subdivision of cerebral cortical dendrites, and in the size of neuronal nuclei and somata in lead-treated rats. Press (1977c) noticed reduction in the size of the external granular and molecular layers of the cerebellum. Brains appeared normal until the fifth day of treatment, when there was an increase in numbers of cell degeneration in the external granular layer. Mitotic activity in this layer was significantly reduced by day 10. The result of these changes was a net loss of cells in the external granular layer. Disruption of cerebellar cortical layers varied and was greatest in those rats which were older, or showed the most severe evidence of encephalopathy.

Several authors have reported Purkinje cell damage. Thomas and Thomas (1974) found that this occurred in patches throughout the cerebellum. Affected cells appeared shrunken, with scalloped borders, pyknotic nuclei, and vacuolated cytoplasm. Electron microscopic changes were reported, including mitochondrial swelling and fragmentation of cristae. Press (1977c) noticed that Purkinje cell development was retarded after 5 days of lead consumption. At day 10, many of these cells possessed features typical of 5–8 day old Purkinje cells. Dendritic trees in the experimental cells were abnormally reduced in height and complexity of branching. In addition, there were changes in synaptogenesis: the perisomatic processes of the lead-intoxicated animals had climbing fibre synapses, rather than the basket cell boutons seen in the more highly developed Purkinje cells of control rats, and also lacked glial investment. A reduction in cerebellar connectivity was also reported by McConnell and Berry (1979). They found that although the Purkinje cell bodies were abnormally large in lead-fed animals (17.5%), the dendritic tree was greatly reduced in overall size. This decrease was due to reductions in segment frequency and distal segment length, rather than in dendritic density, which was normal. Only slight reduction was found in granular cell numbers, suggesting that metabolic changes within the Purkinje cells, rather than outside factors (e.g. parallel fibre alterations) were the cause of the decreases in the network. These changes seemed to result directly from lead exposure, and appeared not to be related to undernutrition. However, in a previous paper (McConnell and Berry, 1978), these authors had reported neuronal effects occurring after neonatal nutritional restriction, including some rather similar Purkinje cell alterations. The possibility of a nutritional effect in the lead-treated animals at the dose employed cannot be entirely dismissed.

In discussion of possible lead effects on the cerebellum, the findings of Klein and Koch (1981) may be relevant. These authors reported that rats dosed intraperitoneally with 5 and 7.5 mg kg^{-1} lead on days 1–10 had more lead in all areas (blood, liver, cerebral cortex, cerebellum and brainstem) than those dosed from days 11–20. In the CNS, the highest concentration of lead was found in the cerebellum at all ages and dose groups, and even untreated day old rats had twice the amount found anywhere else. The cerebellum is the last CNS region to develop, and its structure is not normally completed until well into postnatal life. In many experimental systems lead is administered to maternal rats at parturition, and thus postnatally proliferating areas will be especially affected. This is perhaps why the cerebellum is so severely affected by lead. Observations, including those of Vistica and Ahrens (1977), reported a delay in cerebellar capillary maturation with lead treatment. It would appear then, that the cerebellum may be additionally vulnerable to the effects of lead in early postnatal life by virtue of an imperfectly differentiated blood–brain barrier during the dosing period. Whether it is sensitive to lead by virtue of any other mechanism is still not clear.

Lead was found to be selectively accumulated in the hippocampus, with a concentration seven times that in the entire half brain (ppm dry weight) in mature rats consuming ordinary food pellets and tap water. When adult rats were dosed with 20 ppm lead acetate in drinking water, brains retained lead in similar proportions, and when wet weights were analysed, the hippocampus was found to contain 50% of the total lead in the brain (Fjerdingstad et al., 1974b). In a later paper, a regional analysis of the lead distribution in the hippocampus showed that lead was present in a very high concentration (more than 60 ppm) in the dentate hilus of normal rats. Lead concentration was also high in the granule and molecular layers of the dentate fascia and CA3 region of the hippocampus (about 40 ppm); while in the CA1 region the concentration was only slightly above the hippocampal average of 20 ppm (Danscher et al., 1976).

Experimental studies on the hippocampus suggest a special vulnerability to lead. Louis-Ferdinand (1978) administered 7.5 mg/kg lead acetate via intraperitoneal injection daily from birth to postnatal day 10 and reported a decrease in thickness of the hippocampal cell layer. This deficit was especially evident in the dentate gyrus, in which the granular cell layer showed a lowered neuronal density and prominent nucleoli, and within the hippocampal formation, where the pyramidal cell layer exhibited reduction in thickness. Although this study did not state blood lead values, there was no

117

impairment of weight gain or overt symptoms of toxicity, which suggests the possibility of a specific lead effect. However, the high dose of lead administered might not have caused generalized toxicity merely because the duration of exposure was so short.

Histopathological study of the hippocampus shows other lead-related changes. The dendritic development of hippocampal dentate granule cells was decreased in lead-treated rats, as shown by reduced length of the dendritic field and a reduction of the number of dendritic branches at 160 μm from the cell body; although an increase in branching was reported at 20 μm (Petit and Alfano, 1982). Mossy fibre development has been reported to be affected by lead. Alfano *et al.* (1982) reported several parameters of hippocampal structure to be altered on postnatal days 25 and 60 in offspring of maternal rats dosed with 4% lead carbonate. The maximal width of the hippocampus, the length of the dentate gyrus, and the overall length of the mossy fibre pathway were reduced in size. The spine density on pyramidal cells has also been reported to be decreased (Kiraly and Jones, 1982). The suggestion that lead interferes with the proliferation of hippocampal mossy fibres is supported by Campbell *et al.* (1982), who report a significant decrease in the number of type 1 mossy fibre terminals in the hippocampus of lead-dosed rats. However, these effects may also in part be due to impaired nutrition as animals dosed with 4% lead carbonate are significantly underweight, and hippocampal structure has been shown to be affected by nutritional status (Lewis *et al.*, 1979).

Dendritic alterations were found in the neocortex of lead-treated rats (Petit and LeBoutillier, 1979). Neocortical thickness was reduced, and dendritic trees of the giant pyramidal cells of layer V were underdeveloped. In addition, neocortical synapses were decreased in numerical density, although not in length, width, or other parameters. These authors argued that although there were weight differences between control and lead-dosed animals, the reported synaptic changes differed from those found in undernourished animals. However, significant reductions in cortical dimensions have been reported in undernourished rats (Clark *et al.*, 1973).

Averill and Needleman (1980) reported changes in cortical synaptogenesis with doses as low as 1% lead carbonate. This study used pair-fed controls and showed decrease in the numerical density of synapses in both the lead-treated and undernourished rats. Electron microscope examination revealed normal synaptic development and an absence of degeneration. Although lead-treated animals showed a significantly greater reduction than pair-fed controls at 60 days, it would appear that the effects cited here were strongly linked with undernutrition. Significant decrease in the numerical density of synapses and synapses/neurons was also shown by Krigman *et al.* (1980). Changes persisted to 750 days of age, suggesting that lead was producing a permanent effect. The dose of lead used (200 mg/kg p.o. to neonates daily on days 3–30) would probably produce underweight animals. The persistence of synaptic changes in both these studies gives support to the notion that developmental changes have important repercussions in later life. Whether or not such changes can be attributed directly to lead is another matter.

It can be seen that high levels of lead severely affect the developing nervous system. The cerebellum is most sensitive, and vascular disruption is a primary consequence of high levels of circulating lead in early postnatal development. This vascular disruption 'cascades' onto astroglia, and produces oedema and haemorrhage. Further gliogenesis is probably reactive and stems from non-specific parenchymatous injury, rather than directly from lead. Reported neuronal changes in lead-treated animals (at high doses) are also likely to be and can be attributed to vascular damage, producing anoxia or related metabolic effects indirectly induced. Recently, the hippocampus has become an area of interest in lead neurotoxicity studies, owing to its apparently avid sequestration of lead. The significance of the reported findings will not be clear until definitive studies are carried out using doses of lead that are insufficient to produce systemic toxicity.

LOW DOSE STUDIES

Within the last few years emphasis has shifted from postnatal studies of acute lead intoxication to chronic low dose experimentation. Different areas of the CNS, especially the hippocampus and cerebellum, have been shown to be affected by low levels of lead exposure.

Further studies of the hippocampus have since been undertaken, prompted by the similarity between behavioural aberrations in lead-exposed rats and those with artificially induced damage to both the mature and the developing hippocampus. Minsker et al. (1979; 1982), reported that behaviour in several test systems appeared unaffected by lead exposure in utero or postpartum. However, the fact that lead appears to be selectively concentrated in the hippocampus, and the work reported by Petit and workers above, suggests that lead may be capable of altering hippocampal function directly.

Other brain areas have also been investigated morphologically in animals treated by McCauley et al. (1979), using low concentrations of 50 ppm and 200 ppm Pb as the chloride in the drinking water from 2 weeks before breeding till weaning, found an apparent reduction in the number of measurable synaptic figures (i.e. those including presynaptic dense projections) in the cerebral cortex. A less mature distribution of synapses, independent of actual synaptic density, was also reported. A later study (McCauley et al., 1982) used 30 ppm and 200 ppm Pb (also as the chloride from before breeding till weaning). The normal fourfold increase in synaptic counts in parietal cortex between 11 and 21 days of age was decreased by lead, with a maximal depression at 15 days. After this age there was a catch up, and at 21 days there were no differences in counts. By varying the periods of lead administration to prenatal, perinatal and postnatal dosage, these synaptic differences were found to be largely due to prenatal exposure.

Changes in the retina have also been associated with chronic low level lead exposure, also at 300 ppm and 1000 ppm Pb. These include necrosis of photo-receptor inner segments, perikarya and synaptic endings, Mueller cells, and occasional ganglion cells (Santos-Anderson et al., 1980).

In a preliminary study (Carmichael et al., 1982), rats dosed throughout gestation and lactation with 300 ppm and 1000 ppm lead showed cellular changes in some CNS areas known to be proliferating postnatally. The dentate fascia of the hippocampus was significantly reduced in thickness at days 12 and 16, as was the mitotic index of neuronal precursors in the external granular layer of the cerebellum. The mitotic index of the subependymal layer (an area of gliogenesis) was unchanged in lead-treated rats at all ages and in both dose groups. The fact that two discrete populations of neuronal cells appear to show changes, whereas a population of gliogenic cells does not, seems to suggest a specific neuronal interaction with lead. A later study, looking at the dentate fascia alone, reported that the decrease in cell acqui-sition was not seen at 4 days, but was present in the 300 ppm Pb-dosed rats at 9 days. At 21 days, the differences were not seen, nor were they observed in the adult (Winder et al., 1983c). This suggested that lead caused a transient delay in acquisition in the granule cell layer, but by weaning, a 'catch up' had occurred.

In conclusion, data from low level studies, in which vascular change and undernutrition do not occur, suggest the possibility of specific lead effects on neuronal populations in the forebrain (notably the hippocampus) and cerebellum. To what extent such effects are related to the suspected human problem of subclinical lead neurotoxicity may only emerge after prolonged and intensive study.

SUMMARY

The effect of lead on the peripheral nervous system appears minimal. Effects have been reported, notably on Schwann cells and possibly on axons. However, the lead loads used to produce these effects are very high, and the likelihood of a specific lead effect is small.

Blood lead levels associated with encephalopathy in the rat range from 570 μg Pb/100 ml (LeFauconnier et al., 1980) and 610 μg Pb/100 ml (Press, 1977a) to 1297 μg Pb/100 ml (Alfano and Petit, 1981). Exceptionally, using 4% lead acetate in the maternal diet from parturition, McConnell and Berry (1979) found lesions of encephalopathic type in rats with blood lead levels reported at 258 μg Pb/100 ml.

In rats that do not develop vascular brain lesions with exposure to lead, lower blood lead levels are generally found. Rats with blood levels of 340 and 650 μg Pb/100 ml (LeFauconnier et al., 1980), 331 μg Pb/100 ml (Alfano and Petit, 1981) and between 40 and 194 μg Pb/100 ml (Carmichael et al., 1981) showed no evidence of blood vessel or blood–brain barrier changes. At these dose levels, neuropathological disturbances may be seen (see e.g. Louis-Ferdinand et al., 1978, where it is likely that a blood lead level in the neighbourhood of 250 μg Pb/100 ml was reached). However, the study of Carmichael et al. (1981) showed that with levels above 200 μg Pb/100 ml, systemic nutritional defects were present. LeFauconnier et al. (1980) found significantly reduced neonate weights at 21 days with a blood lead level of 340 μg Pb/100 ml. Since undernutrition itself has powerful effects on brain development and is capable of introducing a wide range of neuropathological changes (see Balazs et al., 1979), it could be argued that only with low levels of exposure, associated with blood lead levels below 200 μg Pb/100 ml, can lesions in the brain be directly related to the toxic metal.

To summarize, a review of published data suggests that lead encephalopathy in the rat is generally found at blood levels above about 600 μg Pb/100 ml and up to 1300 μg Pb/100 ml, and is generally absent with levels below 600 μg Pb/100 ml. Between 200 and 600 μg Pb/100 ml, changes found in the rat brain may possibly derive in part from secondary nutritional disorders as well as from lead. Below 200 μg Pb/100 ml, changes may be seen in the lead-exposed rat nervous system and these are more likely to be due to primary effects of lead.

It was earlier noted that human childhood lead encephalopathy is found with blood lead levels in the range 100–800 μg Pb/100 ml. It is a widely held view that animal models of human lead encephalopathy are of limited value (perhaps due to early experiments on adult animals, which, like man, show relatively little neurological disturbance even with prolonged and intensive dosage). However, this chapter shows that with lead exposure during development, blood levels of the same order of magnitude are associated both in man and rat with comparable encephalopathic lesions.

At lower exposures, the effects of lead are more difficult to examine. It appears that data from recent studies on developing rats with low blood levels (up to 100 μg Pb/100 ml) appear to show effects of lead on maturing and differentiated nerve cell populations. The relevance of these changes to human subclinical lead intoxication is not clear. However, the overall correspondence in lead-poisoned man and rat would make further investigation in this area appear necessary.

7
Discussion

The brain is an organ with considerable structural redundancy and this presumably accounts for some of the ability of the CNS to maintain normal function in the presence of significant damage. Compensation for damage may also occur through biochemical and morphological plasticity (e.g. enzyme induction of axonal sprouting). On the other hand, short-term effects are not likely to cause morphological changes at doses which affect behaviour.

Intrinsically, the use of animals in toxicological studies can be useful as the toxin can be manipulated as an independent variable. This is crudely demonstrated in for example, behavioural toxicology. A toxin is given and subsequent behaviour observed for change. However, the key word in studies of this nature is specificity. It can be very difficult to confirm that any given behavioural observation was directly caused by the toxin, and therefore only in carefully controlled experiments can such conclusions be justified.

As well as specificity of effect, there is also specificity of measure. There is little point in investigating changes in a given system if the measurement of response is too insensitive. There are now a number of sophisticated approaches and procedures which can focus on individual aspects of change. These should be incorporated into experimental design. Examples here include the use of morphometric analysis or behavioural paradigms known to be sensitive for functional deficits.

A third problem is that of animal susceptibility. Some toxic agents will induce a similar response at a given dose in all animals tested. Most do not. What tends to happen is that at any dose a proportion of animals will be affected, and threshold effects occur when the proportion of animals affected becomes significant. The term threshold therefore has only limited validity if the gradient of response is very large. This is the case with lead.

Lastly, there are non-specific effects. High doses of a toxin will be toxic. It therefore becomes necessary to separate non-specific compensatory changes from responses that are believed to be directly induced by the toxin. For lead, the most obvious non-specific effect is undernutrition. There are only two ways of controlling for this effect. These are either to use doses of lead too low to cause undernutrition, or to use undernourished controls. Both have their limitations. In lower dose studies, a careful balance has to be made regarding the exposure of lead: if a dose is too high or too prolonged, non-specific effects begin to dominate; if it is too low, any changes observed may be too minor to even approach significance. Studies using undernourished controls only measure the combined effects of lead and undernutrition and will not control for the interaction between the two.

The presence or absence of any neurotoxic effect is therefore based on methodological and statistical (i.e. extrinsic) factors, as well as intrinsic factors such as specificity and sensitivity.

ANIMAL MODELS

The extrapolation of animal toxicological data to man is always tenuous, but for obvious reasons, animal test models are necessarily used. Unfortunately, there is no single animal model in which effects perfectly correlate with toxicity in children; some 'slippage' is bound to occur in comparisons between the results of animal and clinical or human studies.

For lead, the initial justification for animal experimentation was both simple and immediate. The need for a model of lead administration to investigate the pathogenesis of lead encephalopathy was pressing, as in the 1940s–1960s, this was virtually the only symptom of poisoning considered worth studying. The introduction of reliable and reproducible methods was followed by the publication of many studies establishing the basis of the neuropathology of lead. However, this work had many limitations and later workers tried to improve and refine the methodologies in an attempt to make them more relevant to conditions of human exposure. These in turn, produced further questions regarding the less apparent consequences of elevated lead exposure.

As the nature of the problem of lead intoxication changed, so did the justification for continued experimentation. Studies on encephalopathy were slowly replaced by others designed to investigate the less specific or even debatable aspects of the effects of lower, less toxic levels of lead. The most obvious example of this type was the attempted correlation between lead and hyperactivity. The lead-intoxicated rodent has been used as a reliable animal model for minimal brain dysfunction (MBD), a syndrome commonly known as hyperactivity in children. The suggestion is based primarily on existing information regarding a variety of factors. Specifically, these include the measured neurological sequelae of lead toxicity in mice and rats; the similarity in a paradoxical response of both MBD patients and lead-treated rats to pharmacological agents; and the clinical observation of a possible correlation in young children between the incidence of MBD, and of subclinical toxicity in these same children as a result of exposure.

The investigation into the possible behavioural effects of lead at low doses has produced an enormous literature. All disciplines and approaches have been used, and the numerous different effects reported covered a whole range of neurobiological function. However, the approach to these studies was essentially the same as the original encephalopathy reports. A dose of lead was given that would produce an effect. Non-specific complications were noted, but rarely considered important. Few reports attempted to characterize the model of lead administration employed, and the myriad effects reported were based on a wide range of doses and durations of exposure.

This approach has been changing in recent years. Models of lead administration are being used that have been characterized, or have been shown to be in the clinical range. This has been a major step forward to the understanding not only of the toxic effects of lead, but how this toxicity may be relevant to man.

BEHAVIOUR

It is the function of the CNS that is monitored by behaviour, and as the net sum of neurological processes behaviour therefore is likely to be the most sensitive measure of response to insult. Function may also be the easiest phenomenon to measure in the intact organism, particularly if that organism is man, for whom invasive techniques are often ethically dubious or impractical.

However, the discipline of behavioural toxicology, for all its recent methodological and interpretative improvements, is still a science in its infancy, and a healthy if somewhat unjustified prejudice exists against some of the conclusions made from behavioural studies. This is probably due to the fact that behaviour is a very sensitive to extrinsic (or indeed, intrinsic) influences, and often it may be difficult to separate the two. Another reason is the use of a lexicon that is often difficult to comprehend. Examples from this lexicon include terms such as paradigm, acquisition or performance. Paradigm, when used in this context, is a behavioural test type. Often there may be differences between behavioural apparatus in different locations, but the measured behaviour is the same. An example of this is spontaneous alternation, which can be measured in a T-maze, but also E- or Y-mazes. Acquisition is a curious term which is used extensively in experimental psychology. It is defined as the state when a experimental animal has established some form of response in which it can solve a given experimental situation. This could be called learned or conditioned behaviour. However, it is not possible to conclude that an animal has learned a maze merely because it can solve it. An example of this can be shown in radial mazes, where an animal merely turns left (or, for example, third left) when it leaves a visited arm. To all intents and purposes it may be concluded that the

125

animal has learned the maze, as it will reach any defined criterion rapidly and easily. It can be seen that the animal has not in fact learned anything about the apparatus, but has solved it by responding algorithmically. As it is not possible to discover how the animal is responding in the maze, it therefore is not possible to conclude that the animal is learning the maze. This being the case, a term has to be found that describes the fact that an animal has solved a maze, probably by learning the maze, but taking into account the possibility that the animal has not. This eponym, for all its limitations, is called acquisition. Performance within a maze, can be measured in a variety of ways. Obviously time (latency) and the number of errors are the easist to envisage. In more sophisticated paradigms (e.g. the Hebb–Williams maze) errors are graded. In some instances the terms acquisition and performance may even blur together, especially during the initial stages of an experiment.

The effects of lead on behaviour were, as was seen in Chapter 3, relatively diverse. Initially global measures of activity or emotionality were measured and effects were reported at quite moderate to high exposures of lead. The most notable of these was hyperactivity, thus producing a model for testing childhood hyperactivity, already linked with lead. However, the difference in lead loads between exposed animals and children was large. In some animal models undernutrition was present, itself a probable causal agent in hyperactivity. Low dose studies do not seem to find any effect of lead on activity parameters, although it is possible that some aspects of complex behaviour (such as aggressiveness or social behaviour) may be affected.

Learning studies proved slightly more successful. Differences were seen between control and lead-dosed animals in all manner of learning paradigms. From these, it appears that the harder the task, the more likely lead exposure will cause a learning deficit. Also, improvements in experimental animal psychology were incorporated into the battery of tests available to neuro-toxicologists. The most obvious example of this is the radial arm maze. Over the last 5 years, this piece of apparatus has revolutionized animal psychology. The main reason for this is that the behaviour measured in this apparatus can be correlated to an altered morphology; in this case, the hippocampus. From this it can be assumed that animals showing an altered behaviour may have a change in hippocampal morphology. Of course, if such a structural change is absent, then it is still possible to conclude that the anatomical measure employed is not sensitive enough.

The hypothesis that the behavioural effects of lead are mediated via a lead–hippocampus interaction has received much attention. Indeed, the evidence for the possible involvement of the hippocampus in lead intoxi-cation is quite large, although it is mainly of an indirect nature. From these results, it is possible to assume that lead may cause an alteration in hippo-campal function. However, some of the reported behavioural effects of lead do not correlate well with a hippocampal dysfunction, therefore it is also possible to conclude that the behavioural effects of lead cause a wide variety of effects, part of which may be due to hippocampal damage. Again, these conclusions should be supported by morphological evidence. This evidence is not forthcoming. The correlation of any behavioural change with an altered morphology is an almost insurmountable task, and although it is possible to argue that a behavioural change with no accompanying neuro-biological change is questionable, it is pertinent to ask when objectivity ends and prejudice begins.

NEUROCHEMISTRY

The ultimate unifying objective of neurochemistry is the elucidation of the basic molecular and biochemical mechanisms underlying neural function, or associated with alteration of these functions. As a discipline, it is caught between behaviour, where often it is possible to ascribe a neurochemical basis to a given behaviour (e.g. by pharmacological challenge), and morphology, where a structural change may be preceded by a neurochemical alteration.

The reported neurochemical effects of lead are, as has been seen, enormous. They can be broadly summarized by neurotransmitter system: cholinergic impairment at relatively high levels (usually in the presence of non-specific effects), dopaminergic effects at low levels, and impairment at higher levels (in the presence of obvious non-specific effects), GABAergic effects at levels where a complication from the interaction with D-amino laevulinic acid also occurs. There is also a mixed bag of effects on enkephalin, adenyl cyclase, 5-hydroxytrypamine and other putative aminergic neurotransmitters.

In all, the reported neurochemical effects tend to confuse, rather than aid comprehension of the neurotoxicity of lead. Some effects appear to be seen in the catecholaminergic systems at relatively low doses of lead. As lead-induced changes in other major neurotransmitter systems are generally seen at moderate to high exposures, the question of what is a relevant or justifiable level of lead must again be asked.

NEUROPATHOLOGY

The importance of recognizing the dynamics of pathological processes by using altered morphology (i.e. pathology) in defining cellular malfunction is obvious. In terms of the nervous system, the number of cellular elements which must be considered and identified makes neuropathology appear unduly complex.

However, at the level of the cell, neuropathology is not unique in that neural elements undergo necrosis, injury and repair comparable to cells in systemic tissues. Neuronal injury may be the result of a direct effect on the soma, or secondary to axon injury. Neurones can also be affected by a reduced afferent input and undergo transynaptic degeneration or they may result from damage of supportive glial or vascular processes.

Descriptive pathology therefore has many limitations, not the least being the fact that all pathological lesions (except those that stem from trauma) must be underwritten by a chemical change, and often there is evidence of a neurotoxicity with no discernible pathology. As with other neurobiological disciplines improvements are being made in basic neuropathological approaches, the most important being the application of quantitative (or morphometric) analysis of tissue material. This would be more revealing because the pathological changes may not be in terms of a structural alteration but in a change in the numbers or proportions of critical elements. Morphometric studies are based upon applying stereologic procedures to tissue sections in a defined and co-ordinated manner. The methods may be applied to light and electron microscope material, and will provide information that can be subjected to statistical analysis.

Neuropathological changes are dependent upon recognition of cellular changes and altered ultrastructure. Traditionally, these studies have been based on recognition of qualitative differences; however, few studies have attempted to define alteration in a quantitative manner. Contemporary neuropathological studies, particularly those related to neurotoxicology, should therefore include objective morphometric analyses that can be subjected to statistical analysis. These are necessary not only for demonstrating and defining the subtle undue burdens of toxicants, but for demonstrating the full range of the effects.

In Chapter 6, the study of neuropathological changes in animals induced by lead was reported. Virtually all the reported studies have used doses of lead too high to be of any relevance to lead intoxication in man. This is due to the fact that pathology requires alteration at the cellular level, and the cell must be heavily poisoned before such changes are manifest. The classical encephalopathy of lead in man is produced at high levels of lead, and has been replicated often in experimental animals. The pathogenesis of this encephalopathy has now been established in these animals. The task now facing neuropathologists is whether changes can be found at lower exposures where the damage caused by lead is not so immediately apparent. Some studies have made limited attempts to investigate development, cellular acquisition and synaptogenesis in animals exposed to low levels of lead. As can be seen in Table 10, these are very few in number. The available evidence from these studies suggests there may well be a direct low level effect on neuronal structure, but the limited number of studies indicates that further attention is required before any firm conclusion can be made.

DISCUSSION: OVERVIEW

Lead is a neurotoxin currently the subject of public debate. This is mainly due to the unknown consequences of the practice of adding tetra-alkyl lead additives to petrol, with their consequent dispersal into the atmosphere. The background to this interest has been to establish at what level, if any, lead becomes harmful. As has been seen, lead appears to cause alteration at all levels of neurobiological function. However, interpretation of these effects has been complicated a variety of factors, including levels of dosage, specificity of effect and relevance to man. The evidence for whether lead does or does not have an effect is therefore still not yet concrete. However, the volume of studies showing some form of lead-induced effect make the continued addition of leaded products to petrol inexcusable. Indeed, the perpetuation of this practice appears to be based on procrastination at best or political hypocrisy at worst. The recent report in the UK of a Royal Commission, and the adoption of its findings and suggestions by the UK government (i.e. to remove lead from petrol) is part of a world-wide attempt to limit this practice. This would seem to suggest that interest in this topic will now wane.

In the last 10 years, the neurotoxicological effects of lead have received intense attention, and some of the conclusions reached have been discussed here. Effects of lead have been seen at the behavioural, neurochemical and structural levels in the brain. As it appears that atmospheric lead dispersal will now begin to fall, the emphasis of research into the effects of lead is likely to change again.

8
References

Adams, F. (1844–47). The seven books of Paulus Aeginela (AD 625–690). Translated from the Greek edition of the Aldine Press (Venice, 1528) Vols 1–3, London Sydenham Society

Ahrens, F. A. and Vistica, D. T. (1977). Microvascular effects of lead in the neonatal rat. I. Histochemical and light microscopic studies. *Exp. Mol. Pathol.*, **26**, 129

Aitchinson, L. (1960). *A History of Metals*. Vol 1. (London: MacDonald and Evans)

Alexander, F. W. (1974). The uptake of lead by children in differing environments. *Environmental Health Perspectives*, **7**, 155

Alfano, D. P., LeBoutillier, J. C. and Petit, T. L. (1982). Hippocampal mossy fibre pathway development in normal and postnatally lead-exposed rats. *Exp. Neurol.*, **75**, 308

Alfano, D. P. and Petit, T. L. (1981). Behavioral effects of postnatal lead exposure. Possible relationship to hippocampal dysfunction. *Behav. Neural Biol.*, **32**, 319

Alfano, D. P. and Petit T. L. (1982). Neonatal lead exposure alters the dendritic development of hippocampal dentate granule cells. *Exp. Neurol.*, **75**, 275

Anden, N. E., Robenson, A., Fuxe, K. and Hokfelt, T. (1967). Evidence for dopamine receptor stimulation by apomorphine. *J. Pharm. Pharmacol.*, **19**, 627

Angell, N. F. and Weiss, B. (1982). Operant behavior of rats exposed to lead before or after weaning. *Toxicol. Appl. Pharmacol.*, **63**, 62

Archer, J. (1973). Tests for emotionality in rats and mice: a review. *Animal Behav.*, **21**, 205

vanAssen, F. J. J. (1958). Een geval van loodvergiftiging als oorzaak van aangeboren aafwijkingen bij het nageslacht. *Nederlands Tijdschr. Verloskunde Gynecol.*, **58**, 258

Associated Octel. (1982). *The Lead in Petrol Issue: Reference Literature*. (London: Associated Octel)

Attenburrow, A. A., Campbell, P., Logan, R. W. and Goel, K. M. (1980). Surma and blood lead levels in asian children in Glasgow. *Lancet*, **1**, 323

Aub, J. C., Fairhall, L. T., Minot, A. and Reznikoff, P. (1926). *Lead Poisoning*. (Baltimore: Williams and Wilkins)

Averill, D. M. and Needleman, H. L. (1980). Neonatal lead exposure retards cortical synaptogenesis in the rat. In Needleman, H. L. (ed.). *Low Level Lead Exposure: The Clinical Implications of Current Research*. pp. 201–10. (New York: Raven Press)

Avicenna. See Gruner (1930)

Baker, G. (1767). *An Essay Concerning the Cause of the Endemical Colic of Devonshire*. (London: Hughs) Reprinted by the Delta Omega Society, London (1958)

Baker, J. B. E. (1960). The effects of drugs on the foetus. *Pharmacol. Rev.*, **12**, 37

Balazs, R., Lewis, P. D. and Patel, A. (1979). Nutritional deficiencies and brain development. In Falkner, F. and Tanner, J. M. (eds.). *Human Growth*. Vol 3. pp. 415–80. (London: Balliere Tindal)

Baloh, R. W. (1973). The effects of chronic ingested lead absorption on the nervous system. *Bull. Los Angeles Neurol. Soc.*, **38**, 91

Barlow, J. J., Baruah, J. K. and Davison, A. N. (1977). D-Aminolaevulinic acid dehydratase activity and focal brain haemorrhages in lead-treated rats. *Acta Neuropathol. (Berlin)*, **39**, 219

Barltrop, D. (1969). Transfer of lead to the human fetus. In Barltrop, D. and Barland, W. L. (eds.). *Mineral Metabolism in Pediatrics*. pp. 135–51. (Philadelphia: Davis)

Barocas, R. and Weiss, B. (1974). Behavioral assessment of lead intoxication in children. *Env. Health Perspectives*, **7**, 47

Barrett, J. and Livesey, P. J. (1982). The acetic acid component of lead acetate: its effect on rat weight and activity. *Neurobehav. Toxicol. and Teratol.*, **4**, 105

Barry, P. S. I. (1981). Concentrations of lead in the tissues of children. *Br. J. Industrial Med.*, **38**, 61

Barthalmus, G. T., Leander, J. D., McMillan, D. E., Mushak, P. and Krigman, M. R. (1977). Chronic effects of lead on schedule-controlled pigeon behavior. *Toxicol. Appl. Pharmacol.*, **42**, 271

Baumeister, A. A. (1982). An investigation of the effects of chronic low-level prenatal lead exposure on behavioral development, general activity and learning in neonatal and adult rats. *Dissertation Abstr. Int.*, **42**, 4592 (abstract)

Beattie, A. D., Moore, M. R., Goldberg, A., Finlayson, M. J. W., Mackie, E. M., Graham, J. F., Main, J. C., McClaren, D. A., Murdoch, R. M. and Stewart, G. T. (1975). Role of chronic low level exposure in the aetiology of mental retardation. *Lancet*, **1**, 589

Benignus, V. A., Otto, D. A., Muller, K. E. and Seiple, K. J. Effects of age and body burden on CNS function in young children. II. EEG spectra. *Electroencephalogr. Clin. Neurophysiol.*, **52**, 240

Betts, P. R., Astley, R. and Raine, D. N. (1973). Lead intoxication in Birmingham. *Br. Med. J.*, **1**, 402

Bible. *The New English Bible*. (Oxford: Oxford University Press)

Blackman, S. S. (1936). Intranuclear inclusion bodies in the kidney and liver caused by lead poisoning. *Bull. Johns Hopkins Hosp.*, **61**, 1

Blackman, S. S. (1937). The lesions of lead encephalitis in children. *Bull. Johns Hopkins Hosp.*, **61**, 1

Blair, J. A., Hilburn, M. E., Leeming, R. J., McIntosh, M. J. and Moore, M. R. (1982). Lead and tetrahydrobiopterin metabolism: possible effects on IQ. *Lancet*, **1**, 964

Bondy, S. C., Harrington, M. E., Anderson, C. L. and Prasad, K. N. (1979). The effect of low concentrations of an organic lead concentration on the transport and release of putative transmitters. *Toxicol. Lett.*, **3**, 35

Bornschein, R. L., Fox, D. A. and Michaelson, I. A. Estimation of daily exposure in neonatal rats receiving lead via dam's milk. *Toxicol. Appl. Pharmacol.,* **40,** 577

Bornschein, R., Pearson, D. and Reiter, L. (1980a) Behavioral effects of moderate lead exposure in children and animal models: Part 1, clinical studies. *Crit. Rev. Toxicol.,* **8,** 43

Bornschein, R., Pearson, D. and Reiter, L. (1980b). Behavioral effects of moderate lead exposure in children and animal models. Part 2: animal models. *Crit. Rev. Toxicol.,* **8,** 101

Bouldin, T. W., Mushak, P., O'Tauma, L. A. and Krigman, M. R. (1975). Blood-brain dysfunction in acute lead encephalopathy: A reappraisal. *Env. Health Perspectives,* **12,** 81

Bradley, J. E., Powell, A. E., Nierman, W., McGrady, K. R. and Kaplan, E. (1956). The incidence of abnormal blood levels of lead in a metropolitan pediatric clinic with the observation on value of coproporphyrinuria as a screening test. *J. Pediatr.,* **49,** 1

Brady, K., Herrera, Y. and Zenick, H. (1975). Influence of parental lead exposure on subsequent learning ability of offspring. *Pharmacol. Biochem. Behav.,* **3,** 561

Brashear, C. W., Kopp, V. J. and Krigman, M. R. (1978). Effect of lead on the developing peripheral nervous system. *J. Neuropathy Exp. Neurol.,* **37,** 414

Brennan, M. J. W. and Cantrill, R. C. (1979). D-Aminolaevulinic acid is a potent agonist for GABA autoreceptors. *Nature (London),* **280,** 514

Brown, D. R. Neonatal lead exposure in the rat: decreased learning as a function of age and blood lead concentrations. *Toxicol. Appl. Pharmacol.,* **32,** 628

Bull, R. J., Lutkenhoff, S. D., McCarty, G. E. and Miller, R. G. (1979). Delays in the postnatal increase of cerebral cytochrome concentrations in lead-exposed rats. *Neuropharmacol.,* **18,** 83

Burchfiel, J., Duffy, F., Bartels, P. H. and Needleman, H. L. (1980). Combined discriminating power of quantitative electroencephalography and neuropsychologic measures in evaluating CNS effects of lead at low levels. In Needleman, H. L. (ed.). *Low Level Lead Exposure: the Clinical Implications of Current Research.* pp. 75–89. (New York: Raven Press)

de la Burde, B. (1972). Development of children with elevated blood lead levels. *J. Pediatr.,* **81,** 629

de la Burde, B. and Choate, M. (1975). Early asymptomatic lead exposure and development at school age. *J. Pediatr.,* **87,** 638

Burright, R. B., Donovick, P. J., Michels, K., Fanelli, R. J. and Dolinsky, Z. (1982). Effect of amphetamine and cocaine on seizure activity in lead-treated mice. *Pharmacol. Biochem. Behav.,* **16,** 631

Burton, H. On a remarkable effect upon the human gums produced by the absorption of lead. *Medicine and Chiropody Transactions,* **26,** 63–79 (1840). Also: A remarkable effect on the gums, produced by the slow introduction of lead ions into the human body. *Lancet,* **37,** 665–7 (1839–1840)

Bushnell, P. J. and Bowman, R. E. (1979). Reversal learning deficit in young monkeys exposed to lead. *Pharmacol. Biochem. Behav.,* **10,** 733

Byers, R. K. (1959). Lead poisoning: a review and report of 45 cases. *Pediatrics,* **23,** 585

Cadwalder, T. (1745). *An essay on the West India Dry Gripes . . . an extraordinary case of physick.* (Philadelphia: Benjamin Franklin)

Cahill, D. F., Reiter, L. W., Santolucito, J. A., Rehnberg, G. I., Ash, M. E., Favor, M. J., Bursian, S. J., Wright, J. F. and Laskey, J. W. (1976). Biological assessment of continuous exposure to tritium and lead in the rat. pp. 2–65. In *Biological Effects of Low Level Radiation*. (Vienna: International Atomic Energy Agency)

Campbell, J. B., Woolley, D. E., Vijayan, V. K. and Overmann, S. R. (1982). Morphometric effects of postnatal lead exposure on hippocampal development of the 15-day-old rat. *Dev. Brain Res.*, **3**, 595

Carmichael, N. G., Winder, C. and Lewis, P. D. (1981). Dose response relationships during perinatal lead administration in the rat: a model for the study of lead effects on brain development. *Toxicology*, **21**, 117

Carmichael, N. G., Winder, C. and Lewis, P. D. (1982). Effects of chronic low level intake on the developing rat brain: definition of an experimental system with preliminary findings. *Neuropathol. Appl. Neurobiol.*, **8**, 240

Carpenter, S. J. (1974). Placental transfer of lead. *Env. Health Perspectives*, **7**, 129

Carroll, P. T., Silbergeld, E. K. and Goldberg, A. M. (1977). Alteration of central cholinergic function by chronic lead acetate exposure. *Biochem. Pharmacol.*, **26**, 397

Carson, T. L., van Gelder, G. A., Karas, G. C. and Buck, W. B. (1974). Slowed learning in lambs prenatally exposed to lead. *Arch. Env. Health*, **29**, 154

Caruso, V., Zenick, H. and Michaelson, I. A. (1981). Attenuated behavioral response to D-amphetamine in male and female lead-exposed rats. *Fed. Proc.*, **40**, 700

C.E.C. (1980). *Official Journal of the European Communities*, **L105**, 10 (Council of the European Communities)

Chamberlain, M. J. and Massey, P. M. O. (1972). Mild lead poisoning with an excessively high blood lead. *Br. J. Industrial Med.*, **29**, 458

Charney, E., Satre, J. and Coulter, M. (1980). Increased lead absorption in inner city children: where does the lead come from? *Pediatrics*, **65**, 226

Chisholm Jnr., J. J. (1965). Chronic lead intoxication in children. *Dev. Med. Child Neurol.*, **7**, 529

Chisholm Jnr., J. J. (1972). Development of children with elevated blood lead levels. *J. Pediatr.*, **81**, 628

Chisholm Jnr., J. J. and O'Hara, D. M. (eds.). (1982). *Lead Absorption in Children: Management, Clinical and Environmental Aspects*. (Baltimore: Urban and Schwarzenberg)

Cikrt, M. (1972). Bilary excretion of [203]Hg, [64]Ca, [52]Mn and ^{210}Pb in the rat. *Br. J. Industrial Med.*, **29**, 74

Citois, F. (1616). *De novo poulari apud Pictones dolore colico biloiso Diatriba*. (Poitiers: Mesnier) *See* Major, R. H. (1965)

CLEAR (1983). *Lead versus Health*. Rutter, M. and Russell Jones, R. (eds.). (Chichester: Wiley and sons)

Clark, G. M., Zamenhof, S., van Marthens, E., Grauel, L. and Kruger, L. (1973). The effect of prenatal malnutrition on dimensions of cerebral cortex. *Brain Res.*, **54**, 397

Clarke, R. M. and Hardy, R. N. (1969). An analysis of the mechanism of cessation of macromolecular substances by the intestine of the young rat (closure). *J. Physiol.*, **204**, 127

REFERENCES

Clasen, R. A., Hartman, J. F., Starr, A. J., Coogan, P. S., Pandolfi, S., Laing, I., Becker, R. and Hass, G. M. (1974). Electron microscope and chemical studies of the vascular changes and edema of lead encephalopathy: A comparative study of the human and experimental disease. *Am. J. Pathol.,* **74,** 215

Collins, M. F., Hrdina, P. D., Whittle, E. and Singal, R. L. (1982). Lead in blood and brain regions of rats chronically exposed to low levels of the metal. *Toxicol. Appl. Pharmacol.,* **65,** 314

Conservation Society (1979). Toxic effects of environmental lead. *Proceedings of the symposium held at the Zoological Society of London,* May 1979. (London: The Conservation Society)

Conservation Society (1981). Bryce-Smith, D. and Stephens, R. (eds.). *Lead or Health.* (London: Conservation Society)

Cooper, G. P. and Sigwart, C. D. (1980). Neurophysiological effects of lead. In Singhal, R. L. and Thomas, J. A. (eds.). *Lead Toxicity.* pp. 401–23. (Baltimore: Urban and Schwarzenberg)

Cooper, G. P. and Steinberg, D. (1977). Effects of cadmium and lead on adrenergic transmission in the rabbit. *Am. J. Physiol.,* **232,** C128-C131

Cory-Slechta, D. A., Bissen, S. T., Young, A. M. and Thompson, T. (1981). Chronic postweaning lead exposure and response duration performance. *Toxicol. Appl. Pharmacol.,* **60,** 78

Cory-Slechta, D. A. and Thompson, T. (1979). Behavioral toxicity of chronic post-weaning lead exposure in the rat. *Toxicol. Appl. Pharmacol.,* **47,** 151

Cramer, M. B., Johnson, T. D. and Clarke, D. E. (1980). Low-level lead exposure during growth and development: absence of behavioral and cholinergic neuro-muscular toxicity. *Res. Commun. Substance Abuse,* **1,** 111

Crofton, K. M., Taylor, D. H., Bull, R. J., Sivulka, D. J. and Lutkenhoff, S. D. (1980). Developmental delays in exploration and locomotor activity in male rats exposed to low level lead. *Life Sci.,* **26,** 823

Cutler, M. G. (1977). Effects of exposure to lead on social behaviour in the laboratory mouse. *Psychopharmacology,* **52,** 279

Daines, R. H., Smith, D. W., Feliciano, A. and Tront, J. R. (1972). Air levels of lead inside and outside of homes. *Industrial Med.,* **41,** 26

Dana, S. L. (1848). *Lead Diseases: A Treatise.* (from the French of L. Tanquerel des Planches.) (Boston: Lourell)

Daniels, V. G. (1972). The effect of diet on the time of intestinal macromolecule absorption in the newborn rat intestine. *J. Physiol.,* **226,** 112

Dansher, G., Fjerdingstad, E. J., Fjerdingstad, E. and Fredens, K. (1976). Heavy metal content in subdivisions of the rat hippocampus (zinc, lead and copper). *Brain Res.,* **112,** 442

Danscher, G., Hall, E., Fredens, K., Fjerdingstad, E. and Fjerdingstad, E. J. (1975). Heavy metals in the amygdala of the rat: zinc, lead and copper. *Brain Res.,* **94,** 167

David, O. J. (1974). Association between lower level lead concentrations and hyper-activity in children. *Env. Health Perspectives,* **8,** 17

David, O. (1980). Behavioral and cognitive disorders associated with lead intoxication. In *Electrolytes and Neuropsychiatric Disorders.* pp. 121–42

David, O., Clark, J. and Voeller, K. (1972). Lead and hyperactivity. *Lancet,* **2,** 900

David, O., Grad, G., McGann, B. and Koltun, A. (1982). Mental retardation and 'nontoxic' lead levels. *Am. J. Psychiatr.,* **139,** 806

David, O., Hoffman, S. and McGann, P. J. (1976). Low lead levels and mental retardation. *Lancet,* **2,** 1376

Dawson, E. B., Cravey, W. D., Clark, R. R. and McGanity, W. J. (1969). Effect of trace metals on placental metabolism. *Am. J. Obstetr. Gynecol.,* **103,** 253

Denckla, M. B. and Heilman, K. M. (1979). The syndrome of hyperactivity. In Heilman, K. M. and Valenstein, E. (eds.). *Clinical Neuropsychology.* pp. 574–97. (New York: Oxford University Press)

DeRossett, S. E. (1982). Effects of lead on spontaneous alternation, reactivity, and intercranial self-stimulation. *Dissertation Abstr. Int.,* **42,** 3024-B

Deskin, R., Bursian, S. J. and Edens, F. W. (1980). An investigation into the effects of manganese and other divalent cations on tyrosine hydroxylase activity. *Neurotoxicology,* **2,** 75

Devergie, A. and Hervy, O. (1838). Du cuivre et du plomb. *Annales d'Hygiene Publique et de Medicine Legale,* **20,** 463–5

Dietz, D. D., McMillan, D. E., Grant, L. D. and Kimmel, C. A. (1978). Effects of lead on temporally-spaced responding in rats. *Drug Chem. Toxicol.,* **1,** 401

Dillon, H. K., Wilson, D. J. and Schaffner, W. (1974). Lead concentrations in human milk. *Am. J. Dis. Child.,* **128,** 491

Dioscorides. *See* Goodyer and Gunther (1934)

Dobbing, J. (1968). Vulnerable periods in developing brain. In Davison, A. N. and Dobbing, J. (eds.). *Applied Neurochemistry.* pp. 287–313. (Philadelphia: Davis)

Dolinsky, Z., Fink, E., Burright, R. G. and Donovick, D. J. (1981). The effects of lead, D-amphetamine and time of day on activity levels in the rat. *Pharmacol. Biochem. Behav.,* **14,** 877

Drew, W. G., Kostas, J., McFarland, D. J. and DeRossett, S. E. (1979). Effects of neonatal lead exposure on apomorphine-induced aggression and stereotypy in the rat. *Pharmacology,* **18,** 257

Driscoll, J. W. and Stegner, S. E. (1976). Behavioral effects of chronic lead ingestion on laboratory rats. *Pharmacol. Biochem. Behav.,* **4,** 411

Driscoll, J. W. and Stegner, S. E. (1978). Lead-produced changes in the relative rate of open field activity of laboratory rats. *Pharmacol. Biochem. Behav.,* **8,** 743

Dubas, T. C. and Hrdina, P. D. (1978). Behavioural and neurochemical consequences of neonatal exposure to lead in rats. *J. Env. Pathol. Toxicol.,* **2,** 473

Dubas, T. C., Stevenson, A., Singhal, R. L. and Hrdina, P. D. (1978). Regional alterations of brain biogenic amines in young rats following chronic lead exposure. *Toxicol.,* **9,** 185

Dugandzic, M., Stankovic, B., Milovanovic, L. and Koricanac, Z. (1973). Urinary excretion of 5-hydroxyindole acetic acid in lead-exposed persons. *Ark. Hig. Rada Toxsikol.,* **24,** 37

Dyck, P. J., Windebank, A. J., Low, P. A. and Baumann, W. J. (1980). Blood nerve barrier in rat and cellular mechanisms of lead-induced segmental demyelination. *J. Neuropathol. Exp. Neurol.,* **39,** 700

E.E.C. (1975). *Proceedings of an International Symposium of Recent Advances in the Assessment of the Health Effects of Environmental Pollution.* (Luxemburg: E.E.C.)

Elwood, P. C. Mis-lead by the E.E.C. *Lancet,* **2,** 824

Elwood, W. J., Clayton, B. E., Cox, R. A., Delves, H. T., King, E., Malcolm, D.,

Ratcliffe, J. M. and Taylor, J. F. (1977). Lead in human blood and in the environment near a battery factory. *Br. J. Prevent. Social. Med.*, **31**, 154

Ernhart, C. B., Landa, B. and Schell, N. B. (1981). Subclinical levels of lead and developmental deficit – a multivariate follow-up reassessment. *Pediatrics*, **67**, 911

Ewers, U. and Erbe, R. (1980). Effects of lead, cadmium and mercury on brain adenylate cyclase. *Toxicology*, **16**, 227

Feeley, D. M., Longo, J. F., Cosden, M. A., Zenick, H. and Padich, R. (1979). Detection of the effects of lead exposure by visual evoked response latency. *Physiol. Psychol.*, **7**, 143

Fellowship of Engineering (1981). *Reduction of Lead in the Environment – Energy, Technology and Cost: Report of a Working Party on Lead.* (London: The Fellowship of Engineering)

Ferm, V. H. and Carpenter, S. J. (1967). Developmental malformation resulting from the administration of lead salts. *Exp. Mol. Pathol.*, **7**, 208

Fine, P. R. and Dobin, D. D. (1975). The incidence of elevated blood lead values in an asymptomatic pediatric population residing in a major American industrial state. In *Proceedings of an International Symposium of Recent Advances in the Assessment of the Health Effects of Environmental Pollution.* pp. 1223–31. (Luxemburg: EEC)

Fine, P. R., Thomas, C. W., Suhs, R. H., Colhnberg, R. E. and Flashner, B. A. (1972). Pediatric blood lead levels: a study of 14 Illinois cities of intermediate population. *J. Am. Med. Assoc.*, **221**, 1475

Fjerdingstad, E., Danscher, G. and Fjerdingstad, E. J. (1974a). Zinc content in hippocampus and whole brain of normal rats. *Brain Res.*, **79**, 338

Fjerdingstad, E. J., Danscher, G. and Fjerdingstad, E. (1974b). Hippocampus: selective concentration of lead in the normal rat brain. *Brain Res.*, **80**, 350

Flynn, E. R. (1979). Neurochemical and behavioral effects of prenatal lead ingestion: Brain lead content, brain calcium, activity level and maze learning. *Diss. Abstr. Int.*, **39**, 6183-B

Flynn, J. C., Flynn, E. R. and Patton, J. H. (1979). Effects of pre- and post-natal lead on affective behavior and learning in the rat. In: Test methods for definition of effects of toxic substances on behavior and neuromotor function. *Neurobehav. Toxicol.*, **1**, (Suppl. 1), 93

F.O.E. (Price, B.) (1982). *Lead in Petrol: An Energy Analysis.* (London: Friends of the Earth)

Fowler, B. A., Kimmel, C. A., Woods, J. S., McConnell, E. E. and Grant, L. D. (1980). Chronic low level lead toxicity in the rat. III. An integrated assessment of long term toxicity with special reference to the kidney. *Toxicol. Appl. Pharmacol.*, **56**, 59

Forbes, R. J. (1950). *Metallurgy in Antiquity.* (Leiden: Brill)

Forfar, J. O. and Arneil, G. C. (1973). *Textbook of Paediatrics.* pp. 1728–9. (Edinburgh and London: Churchill Livingstone)

Fox, D. F., Lewkowski, J. P. and Cooper, G. P. (1977). Acute and chronic effects of neonatal lead exposure on development of the visual-evoked response in rats. *Toxicol. Appl. Pharmacol.*, **40**, 449

Fox, D. F., Lewkowski, J. P. and Cooper, G. P. (1979a). Persistent visual cortex excitability alterations produced by neonatal lead exposure. *Neurobehav. Toxicol.*, **1**, 101

Fox, D. A., Overmann, S. R. and Woolley, D. E. (1979b). Neurobehavioral ontogeny of neonatally lead-exposed rats. II. Maximal electroshock seizures in developing and adult rats. *Neurotoxicology*, **1**, 149

Gale, N. H. and Stos-Gale, Z. (1981). Lead and silver in the ancient Aegean. *Sci. Am.*, **224**, 142

Gatzke, H.-D. (1980). The influence of lead on the protein metabolism of the brain. *Anatom. Anzeiger (Jena)*, **148**, 309

Geist, C. R. and Balko, S. W. (1980). Effects of postnatal lead acetate on activity and emotionality in developing laboratory rats. *Bull. Psychonomic Soc.*, **15**, 288

Geist, C. R. and Mattes, B. R. (1979). Behavioral effects of postnatal lead acetate exposure in developing laboratory rats. *Physiol. Psychol.*, **7**, 399

Geist, C. R. and Praed, J. E. (1982). Chronic lead exposure of rats: open-field performance. *Perceptual Motor Skills*, **55**, 487

Gelman, B. G. and Michaelson, I. A. (1979). Neonatal lead toxicity and *in vitro* lipid peroxidation of rat brain. *J. Toxicol. Env. Health*, **5**, 671

Gelman, B. G., Michaelson, I. A. and Bornschein, R. L. (1979). Brain lipofuscin concentration and oxidant defence enzymes in lead-poisoned neonatal rats. *J. Toxicol. Env. Health*, **5**, 683

Gerber, G. B., Maes, J., Gilliavod, N. and Casale, G. (1978). Brain biochemistry of infant mice and rats exposed to lead. *Toxicol. Lett.*, **2**, 51

Gibson, S. L. M., Lam, C. N., McGrae, W. M. and Goldberg, A. (1967). Blood lead levels in normal and mentally deficient children. *Arch. Dis. Child.*, **42**, 573

Gilfallen, S. C. (1966). Lead poisoning and the fall of Rome. *J. Occupational Med.*, **7**, 53

Goldman, D., Hejtmancik, M. R., Williams, B. J. and Ziegler, M. G. (1980). Altered noradrenergic systems in the lead-exposed neonatal rat. *Neurobehav. Toxicol.*, **2**, 337

Goldstein, G. W. and Diamond, I. (1974). Metabolic basis of lead encephalopathy. In Plum, F. (ed.). *Brain Dysfunction in Metabolic Disorders*. pp. 293–304. (New York: Raven Press)

Goldstein, G. W., Asbury, A. K. and Diamond, I. (1974). Pathogenesis of lead encephalopathy. Uptake of lead and reaction of brain capillaries. *Arch. Neurol.*, **31**, 382

Golter, M. and Michaelson, I. A. (1975). Growth, behavior, and brain catecholamines in lead-exposed neonatal rats: a reappraisal. *Science*, **187**, 359

Gombault, A. (1880). Contributions a l'etude de la nervite parenchymateuse subaigue et chronique nervite segmentaile periaxiale. *Arch. Neurol. (Paris)*, **1**, 11

Goodyer (translater) and Gunther (editor) (1934). *The Greek Herbal of Dioscorides*. (London: Oxford University Press)

Gordon, N., King, E. and Mackay, R. I. (1967). Lead absorption in children. *Br. Med. J.*, **2**, 480

Govoni, S., Memo, M., Lucchi, L., Spano, P. F. and Trabucchi, M. (1980). Brain neurotransmitter systems and chronic lead intoxication. *Pharmacol. Res. Commun.*, **12**, 447

Govoni, S., Memo, M., Spano, P. F. and Trabucchi, M. (1979). Chronic lead treatment differentially affects dopamine synthesis in various brain areas. *Toxicology*, **12**, 343

Govoni, S., Montefusco, O., Spano, P. F. and Trabucchi, M. (1978). Effect of chronic lead treatment on brain dopamine synthesis and serum prolactin release in the rat. *Toxicol. Lett.,* **2,** 333

Gowland, W. (1912). The metals in antiquity. *J. Anthropol. Res.,* **42,** 235

Grandjean, P. (1978). Regional distribution of lead in human brains. *Toxicol. Lett.,* **2, 65**

Granick, J. L., Sassa, S. and Kappas, A. (1978). Some biochemical and clinical aspects of lead intoxication. *Adv. Clin. Chem.,* **20,** 287

Grant, L. D., Kimmel, C. A., Martinez-Vargas, C. M. and West, G. L. (1976). Assessment of developmental toxicity associated with chronic lead exposure. *Env. Health Perspectives,* **17,** 290

Grant, L. D., Kimmel, C. A., West, G. L., Martinez-Vargas, C. M. and Howard, J. L. (1980). Chronic low-level lead toxicity in the rat. II. Effects on postnatal physical and behavioral development. *Toxicol. Appl. Pharmacol.,* **56,** 42

Greenberg, M., Jacobziner, H., McLaughlin, M. C., Fuerst, H. T. and Pellitten, O. (1958). A study of pica in relation to lead poisoning. *Pediatrics,* **22,** 756

Greengard, J. (1966). Lead poisoning in childhood: Signs, symptoms current therapy, clinical expressions. *Clin. Pediatr.,* **5,** 269

Griggs, R. C., Sunshine, I., Newill, V. A., Newton, B. W., Buchanan, S. and Rasch, C. A. (1964). Environmental factors in childhood lead poisoning. *J. Am. Med. Assoc.,* **187,** 703

Grisolle, A. (1836). Recherches sur quelques-uns des accidents cerebraux prodiuts par les preparationes saturnines. (Paris)

Gross-Selbeck, E. and Gross-Selbeck, M. (1980). A biphasic effect of lead on operant behaviour of rats induced by different exposure. *Toxicol. Lett. (Special Issue)* **0,** 128 (abstract P8)

Gross-Selbeck, E. and Gross-Selbeck, M. (1981). Changes in operant behavior of rats exposed to lead at the accepted no-effect level. *Clin. Toxicol.,* **18,** 1247

Gruden, N. (1975). Lead and active calcium transfer through the intestinal wall in rats. *Toxicology,* **5,** 163

Gruner, O. C. (1930). *A Treatise on the Canon of Medicine of Avicenna.* (London: Luzac)

Hall, C. S. (1934). Emotional Behavior in the rat: III. Defecation and urination as measures of individual differences in emotionality. *J. Comp. Psychol.,* **18,** 385

Hall, C. S. and Bellechey, E. L. (1932). A study of the rats behavior in a field: a contribution to method in comparitive psychology. *University California Publ. Psychol.,* **6,** 1

Hansard (House of Commons) (1976). Water Supplies (Lead Content). Written Answers. **903,** col 540

Harris, P. and Holley, M. R. (1972). Lead levels in cord blood. *Pediatrics,* **49,** 606

Harry, G. J., Goodrum, J. F., Mushak, P., Krigman, M. R. and Morell, P. (1982). CNS damage resulting from ingestion of organic lead is not directly related to blood lead level. *Fed. Proc.,* **41,** 1561 (abstract 7544)

Hastings, L., Cooper, G. P. and Bornschein, R. L. (1976). The effect of early lead exposure on the behavior of developing rats. *Toxicol. Appl. Pharmacol.,* **37,** 162 (abstract 169)

Hastings, L., Cooper, G. P., Bornschein, R. L. and Michaelson, I. A. (1977).

Behavioral effects of low level neonatal lead exposure. *Pharmacol. Biochem. Behav.*, **7**, 37

Hastings, L., Cooper, G. P., Bornschein, R. L. and Michaelson, I. A. (1979). Behavioral deficits in adult rats following neonatal lead exposure. *Neurobehav. Toxicol.*, **1**, 227

Hebel, J. R., Kinch, D. and Armstrong, E. (1976). Mental capability of children exposed to lead pollution. *Br. J. Preventive Social Med.*, **30**, 170

Hejtmancik, M. R., Dawson, E. B. and Williams, B. J. (1982). Tissue distribution of lead in rat pups nourished by lead poisoned mothers. *J. Toxicol. Env. Health*, **9**, 77

Heroditus (450 BC). *See* de Selincourt (1966)

Hippocrates. *See* Jones, W. H. S. and Withington, E. T. (1923–31)

Hirano, A. and Iwata, M. (1979). Neuropathology of lead intoxication. Handbook of Clinical Intoxication: Intoxications of the Nervous System. Part I (eds. Vinken, P. J. and Bruyn, G. W.) (Amsterdam: Elsevier)

Holtzman, D., Herman, M. M., Hsu, J. S. and Mortell, P. (1980). The pathogenesis of lead encephalopathy: effects of lead carbonate feedings on morphology, lead content and mitochondrial respiration in brains of immature and adult rats. *Virchows Arch. A, Pathol. Anat.*, **387**, 147

Holtzman, D. and Hsu, J. S. (1976). Early effects of inorganic lead on immature rat brain mitochondrial respiration. *Pediatr. Res.*, **10**, 70

Holtzman, D., Hsu, J. S. and Desautel, M. (1981). Absence of effects of lead feedings and growth retardation on mitochondrial and microsomal cytochromes in the developing brain. *Toxicol. Appl. Pharmacol.*, **58**, 48

Hopkins, A. P. and Dayan, A. D. (1974). The pathology of experimental lead encephalopathy in the baboon *(Papio anubis)*. *Br. J. Industrial Med.*, **31**, 128

Horace. Quinti Horatii Flacci: Opera Omnia. *See* Wickham. E. C. (1891)

Hrdina, P. D., Hanin, I. and Dubas, T. C. (1980). Neurochemical correlates of lead toxicity. In Singhal, R. L. and Thomas, J. A. (eds.). *Lead Toxicity*. pp. 273–300. (Baltimore: Urban and Schwarzenberg)

Hrdina, P. D., Peters, D. A. V. and Singhal, R. L. (1976). Effects of chronic exposure to cadmium, lead and mercury of brain biogenic amines in the rat. *Res. Commun. Chem. Pathol. Pharmacol.*, **15**, 483

H.S.M.H.A. (1971). Medical aspects of childhood lead poisoning. *Health Services and Mental Health Administration Health Reports*, **86**, 140

Hsu, J. M. (1981). Lead toxicity as related to glutathione metabolism. *J. Nutr.* **111**, 26

Hunter, D. (1975). *The Diseases of Occupations*. 5th edn. (London: English Universities Press)

Huxham, J. De morbo colico Damnoniensi. London: S. Austen (1739). Also: Observations on the Air and Epidemic Diseases with a Short Dissertation on the Devonshire Colic. London: Henton (1759). *See* Major, R. H. (1965)

Isaacson, R. L. (1974). *The Limbic System*. (New York: Plenum Press)

Jacobson, M. (1978). *Developmental Neurobiology*. 2nd Edn. (New York: Plenum Press)

Jacobziner, H. (1966). Lead poisoning in childhood: epidemiology, manifestations and prevention. *Clin. Pediatr.*, **5**, 277

James, L. F., Lazar, V. A. and Binns, W. (1966). Effects of sublethal doses of certain minerals on pregnant ewes and fetal development. *Am. J. Vet. Res.*, **27**, 132

Jason, K. and Kellogg, C. (1977). Lead effects on behavioral and neurochemical development in rats. *Fed. Proc.*, **36**, 1008 (abstract 3887)

Jason, K. M. and Kellogg, C. K. (1981). Neonatal lead exposure: effects on development of behavior and striatal dopamine neurones. *Pharmacol. Biochem. Behav.*, **15**, 641

Jones, D. L. (1979). Effects of neonatal lead on spontaneous alternation performance in rats. *Soc. Neurosci. Abstr. Psychopharmacol.*, **5**, 650 (abstract 2214)

Jones, W. H. S. and Rackham, H. (1938–63). *Natural History of Pliny the Elder (English translation)*. 10 vols. (London: W. Heinemann) (1938–63)

Jones, W. H. S. and Withington, E. T. (1923–31). *The Works of Hippocrates (English translation)*. (London: W. Heineman)

Jordan, T. C., Cane, S. E. and Howells, K. F. (1981). Deficits in spatial memory performance induced by early undernutrition. *Dev. Psychobiol.*, **14**, 317

Jordan, T. C., Howells, K. F. and Cane, S. E. (1980). Hippocampal and spatial memory deficits resulting from early undernutrition. In DiBenedetta, C., Balazs, R., Gombos, G. and Porcellati, G. (eds.). *A Multidisciplinary Approach to Brain Development. (Proceedings of European Society of Neurochemistry Meeting held at Brindisi, Italy,* April 16–21, 1979. pp. 347–8 (Amsterdam: Elsevier)

Jordan, T. C., Howells, K. F., McNaughton, N. and Heatlie, P. L. (1982). Effects of early undernutrition on hippocampal development and function. *Res. Exp. Med.*, **180**, 201

Karnofsky, D. A. (1965). Drugs as teratogens in animals and man. *Annu. Rev. Pharm.*, **5**, 477

Kehoe, R. A. (1960). The Harben lectures. The metabolism of lead in man in health and disease. *J. R. Inst. Public Health Hyg.* pp. 24, 81–96, 101–20, 129–43 and 177–203

Kehoe, R. A., Thamann, F. and Cholak, J. (1933). On the normal absorption and excretion of lead. *J. Industrial Hyg.*, **15**, 257

Kehoe, R. A., Thamann, F. and Cholak, J. (1934). An appraisal of the lead hazards associated with the distribution and use of gasoline containing tetraethyl lead. Part 1. *J. Industrial Hyg.*, **16**, 100

Kemp, K. and Danscher, G. (1979). Multi-element analysis of the rat hippocampus by proton induced X-ray emission spectroscopy (phospurus, sulphur, chlorine, potassium, calcium, iron, zinc, copper, lead, bromine and rubidium). *Histochemistry*, **59**, 167

Kihara, T., Matsuo, T., Kamimura, M., Yasuda, Y. and Tanimura, T. (1981). Effects of maternal lead acetate exposure on behavioral and reproductive performance of post-weaning rats. *Teratology*, **24**, 31A

Kiraly, E. and Jones, D. G. (1982). Dendritic spine changes in rat hippocampal pyramidal cells after postnatal lead treatment: A Golgi study. *Exp. Neurol.*, **77**, 236

Kishi, R., Ikeda, T., Miyake, H., Uchino, E., Tsuzuki, T. and Inoue, K. (1982). Regional distribution of lead, copper and zinc in suckling and adult brains. *Brain Res.*, **251**, 180

Klein, A. B. and Koch, T. R. (1981). Lead accumulations in brain, blood and liver after low dosing of neonatal rats. *Arch. Toxicol.*, **47**, 257

Kober, T. E. and Cooper, G. P. (1977). Lead competitive calcium dependent synaptic transmission in the bullfrog sympathetic ganglion. *Nature (London)*, **262**, 704

Kochen, J. and Greener, Y. (1977). Brain levels in hemorrhagic lead encephalopathy. *Pediatr. Res.*, **11**, 563 (abstract 1151)

141

Komulainen, H. and Tuomisto, J. (1981). Effect of heavy metals on dopamine, noradrenaline and serotonin uptake and release in rat brain synaptosomes. *Acta Pharmacol. Toxicol.*, **48**, 199

Kostas, J. McFarland, D. J. and Drew, W. G. (1976). Lead-induced hyperactivity. Chronic exposure during the neonatal period in the rat. *Pharmacology*, **14**, 435

Kostas, J., McFarland, D. J. and Drew, W. G. (1978). Lead-induced behavioral disorders in the rat: Effects of amphetamine. *Pharmacology*, **16**, 226

Kostial, K., Simonvic, I. and Pisonic, M. (1971). Lead absorption from the intestine of newborn rats. *Nature (London)*, **233**, 564

Kostial, K. and Vouk, V. B. (1957). Lead ions and synaptic transmission in the superior cervical ganglion of the cat. *Br. J. Pharmacol.*, **12**, 219

Kotok, D. (1972). Development of children with elevated blood lead levels: A controlled study. *J. Pediatr.*, **80**, 57

Kotok, D., Kotok, R. and Heriot, J. T. (1977). Cognitive evaluation of children with elevated blood lead levels. *Am. J. Dis. Child.*, **131**, 791

Krall, A. R., Pesavento, C., Harmon, S. J. and Packer III, R. M. (1972). Elevation of norepinephrine levels and inhibition of mitochondrial oxidative phosphorylation in cerebellum of lead-intoxicated suckling rats. *Fed. Proc.*, **31**, 665 (abstract 2537)

Krall, V., Sachs, H., Rayson, B., Lazar, B., Growe, G. and O'Connell, L. (1980). Effects of lead poisoning on cognitive test performance. *Perceptual Motor Skills*, **50**, 483

Krass, B., Winneke, G. and Kramer, U. (1980). Neurobehavioral and systemic effects in lead-exposed rats after an exposure-free interval of four months duration. *Zentralblatt Bakteriol. I Abt. Original. B.*, **170**, 353

Krehbiel, D., Davis, G. A., Leroy, L. M. and Bowman, R. E. (1976). Absence of hyperactivity in lead-exposed developing rats. *Env. Health Perspectives*, **18**, 147

Krigman, M. R., Bouldin, T. W., Gaynor, J. and Bagnell, C. R. (1980). Effect of lead burdens on synapses in the aging rat caudate nucleus. *Fed. Proc.*, **40**, 78 (abstract 3183)

Krigman, M. R., Butts, S. A., Hogan, E. L. and Shinkman, P. G. (1972). Morphological, neurochemical, and behavior correlates of lead intoxication and undernourishment in developing rats. *Fed. Proc.*, **31**, 665 (abstract 2536)

Krigman, M. R., Druse, M. J., Traylor, T. D., Wilson, M. H., Newell, L. R. and Hogan, E. L. (1974a). Lead encephalopathy in the developing rat: Effect upon myelination. *J. Neuropathol. Exp. Neurol.*, **33**, 58

Krigman, M. R., Druse, M. J., Traylor, T. D., Wilson, M. H., Newell, L. R. and Hogan, E. L. (1974b). Lead encephalopathy in the developing rat: effect on cortical ontogenesis. *J. Neuropathol. Exp. Neurol.*, **33**, 671

Krigman, M. R. and Hogan, E. L. (1974). Effect of lead intoxication on the postnatal growth of the rat nervous system. *Env. Health Perspectives*, **7**, 187

Kubasik, N. P. and Volosin, M. T. (1972). Concentrations of lead in capillary blood of newborns. *Clin. Chem.*, **18**, 1415

Kussmaul, A. and Maier, R. (1872). Zur pathologischen anatomie des chronischen saturnismus. *Deutsche. Arch. Klin. Med.*, **9**, 283

Lamm, S., Cole, B., Glynn, K. and Ullmann, W. (1973). Lead content of milk fed to infants 1971–1972. *N. Engl. J. Med.*, **289**, 574

Lampert, P. F., Garro, F. and Pentschew, A. (1967). Lead encephalopathy in suckling rats: An electron microscopic study. In Klatzo, I. and Seitelberger, F. (eds.). *Symposium on Brain Edema.* pp. 207-23. (New York: Springer)

Lampert, P. F. and Schochet, S. S. (1968). Demyelination and remyelination in lead neuropathy: electron microscope studies. *J. Neuropathol. Exp. Neurol.,* **27,** 527

Lancranajan, I., Popescu, H. I., Gavanesu, O., Klapsh, I. and Serbanescu, M. (1975). Reproductive ability of workmen occupationally exposed to lead. *Toxicol. Appl. Pharmacol.,* **30,** 396

Landrigan, P. J., Whitworth, R. H., Baloh, R. N., Staehling, N. W., Barth, W. F. and Rosenblum, B. F. (1975). Neuropsychological dysfunction in children with low level lead absorption. *Lancet,* **1,** 708

Lang, E. P. and Kunze, F. M. (1948). The penetration of lead through the skin. *J. Industrial Hyg.,* **30,** 256

Lansdowne, R. G., Shepherd, J., Clayton, B. E., Delves, H. T., Graham, R. J. and Turner, W. C. (1974). Blood levels behaviour and intelligence: A population study. *Lancet,* **1,** 538

Lanthorn, T. and Isaacson, R. L. (1978). Effects of chronic lead ingestion in adult rats. *Physiol. Psychol.,* **6,** 93

Lauwerys, R., Buchet, J. P., Roels, H. and Hubermont, G. (1978). Placental transfer of lead, mercury, cadmium and carbon monoxide in women. *Env. Res.,* **15,** 278

Lawther, P. J., Cummings, B. T., McEllison, J. and Biles, B. (1972). Airbourne lead and its uptake by inhalation. In Hepple, P. (ed.). *Lead in the Environment.* pp. 8-28. (Organized by the Institute of Petroleum) (London: Applied Science)

Lazareno, S. and Nahorski, S. R. (1982). Selective labelling of dopamine (D2) receptors in rat striatum by [³H]domperidone but not by [³H]spiperone. *Eur. J. Pharmacol.,* **81,** 273

Leeming, R. J. and Blair, J. A. (1980). The effects of pathological and normal physiological processes on biopterin derivative levels in man. *Clin. Chim. Acta,* **108,** 103

LeFauconnier, J. M., Bernard, G., Mellerio, F., Sebille, A. and Cesarini, E. (1983). Lead distribution in the nervous system of 8 month-old rats intoxicated since birth by lead. (Submitted to *Experientia*)

LeFauconnier, J. M., Lavielle, E., Terrien, N., Bernard, Fournier, E. (1980). Effect of various lead doses on some cerebral capillary functions in the suckling rat. *Toxicol. Appl. Pharmacol.,* **55,** 467

Legge, Sir T. M. and Goadby, Sir K. W. (1912). *Lead Poisoning and Lead Absorption.* (London: E. Arnold)

Lewis, P. D., Patel, A. J. and Balazs, R. (1979). Effect of undernutrition on cell generation in rat hippocampus. *Brain Res.,* **168,** 186

Lin Fu, J. S. (1973). Vulnerability of children to lead exposure and toxicity. *N. Engl. J. Med.,* **289,** 1229

Lin Fu, J. S. (1979). Lead exposure among children - a reassessment. *N. Engl. J. Med.,* **300,** 731

Loch, R. S., Rafales, L. S., Michaelson, I. A. and Bornschein, R. L. (1978). The role of undernutrition in animal models of hyperactivity. *Life Sci.,* **22,** 1963

Louis-Ferdinand, R. T., Brown, D. R., Fiddler, S. F., Daughtrey, W. C. and Klein, A. W. (1978). Morphometric and enzymatic effects of neonatal lead exposure in the rat brain. *Toxicol. Appl. Pharmacol.,* **43,** 351

Lucas, A. and Harris, J. R. (1962). *Ancient Egyptian Materials and Industries*. 4th Edn. (London: E. Arnold)

Lucchi, L., Memo, M., Airaghi, M. L., Spano, P. F. and Trabucchi, M. (1981). Chronic lead treatment induces in rat a specific and differential effect on dopamine receptors in different brain areas. *Brain Res., 213*, 397

Mackie, A. C., Stephens, R., Townshend, A. and Waldron, H. A. (1977). Tooth lead levels in Birmingham children. *Arch. Env. Health, 32*, 178

Martin, A. E., Fairweather, F. A., Buxton, R. St. J. and Roots, L. M. (1975). Recent epidemiologic studies of environmental lead of industrial origin. In *Proceedings of an International Symposium of Recent Advances in the Assessment of the Health Effects of Environmental Pollution*. pp. 1113–20 (Luxemburg: E.E.C.)

Major, R. H. (1965). *Classic Descriptions of Disease*. (Springfield: C. C. Thomas)

McCauley, P. T., Bull, R. J. and Lutkenhoff, S. D. (1979). Association of alterations in energy metabolism with lead-induced delays in rat cerebral cortical development. *Neuropharmacology, 18*, 93

McCauley, P. T., Bull, R. J., Tonti, A. P., Lutkenhoff, S. D., Veister, M. V., Doerger, J. U. and Stober, J. A. (1982). The effect of prenatal and postnatal lead exposure on neocortical synaptogenesis in rat cerebral cortex. *J. Toxicol. Env. Health, 10*, 639

McClain, R. M. and Becker, B. A. (1972). Effects of organolead compounds on rat embryonic and fetal development *Toxicol. Appl. Pharmacol., 21*, 265

McClain, R. M. and Becker, B. A. (1975). Teratogenicity, fetal toxicity and placental transfer of lead nitrate in rats. *Toxicol. Appl. Pharmacol., 31*, 72

McConnell, P. and Berry, M. (1978). The effects of undernutrition on Purkinje cell dendritic growth in the rat. *J. Comp. Neurol., 177*, 159

McConnell, P. and Berry, M. (1979). The effects of postnatal lead exposure on Purkinje cell dendritic development in the rat. *Neuropathol. Appl. Neurobiol., 5*, 115

McCord, C. P. (1953). Lead and lead poisoning in America. Benjamin Franklin and lead poisoning. *Industrial Med., 22*, 393

McIntosh, M. J., Moore, M. R., Blair, J. A., Milburn, M. E. and Leeming, R. J. (1982). Lead and tetrahydrobiopterin metabolism in man and animals. *Med. Res. Soc. Abstr. Clin. Sci., 63*, 44p (abstract 121)

McLaughlin, M. C. (1956). Lead poisoning in children in New York city, 1950–1954: epidemiologic study. *N. York State J. Med., 56*, 3711

Memo, M., Lucchi, L., Spano, P. F. and Trabucchi, M. (1980a). Lack of correlation between the neurochemical and behavioural effects induced by D-amphetamine in chronically lead-treated rats. *Neuropharmacology, 19*, 795

Memo, M., Lucchi, L., Spano, P. F. and Trabucchi, M. (1980b). Effect of lead treatment on gaba-ergic receptor function in rat brain. *Toxicol. Lett., 6*, 427

Memo, M., Lucchi, L., Spano, P. F. and Trabucchi, M. (1980c). Dose-dependent effects of lead on different neurotransmitter systems in various rat brain areas. In Manzo, L. (ed.). *Advances in Toxicology: Proceedings of the International Congress on Neurotoxicology*. (New York: Pergamon)

Memo, M., Lucchi, L., Spano, P. F. and Trabucchi, M. (1981). Dose-dependant and reversible effects of lead on rat dopaminergic system. *Life Sci., 28*, 795

Michaelson, I. A. (1973). Effects of inorganic lead on RNA, DNA and protein content in the developing neonatal rat brain. *Toxicol. Appl. Pharmacol., 26*, 539

Michaelson, I. A. (1980). An appraisal of rodent studies on the behavioral toxicity of lead: the role of nutritional status. In Singhal, R. L. and Thomas, J. A. (eds.). *Lead Toxicity*. pp. 301–65. (Baltimore: Urban and Schwarzenberg)

Michaelson, I. A. and Bradbury, M. (1982). Effect of early inorganic lead exposure on rat blood-brain barrier permeability to tyrosine or choline. *Biochem. Pharmacol.*, **31**, 1881

Michaelson, I. A., Greenland, R. D. and Roth, W. (1974). Increased brain norepinephrine turnover in lead-exposed rats. *Pharmacologist*, **16**, 250 (abstract 340)

Michaelson, I. A. and Sauerhoff, M. W. (1974a). An improved model of lead-induced brain dysfunction in the suckling rat. *Toxicol. Appl. Pharmacol.*, **28**, 88

Michaelson, I. A. and Sauerhoff, M. W. (1974b). Animal models of human disease: Severe and mild lead encephalopathy in the neonatal rat. *Env. Health Perspectives*, **7**, 201

Milar, C. P., Shroeder, S. R., Mushak, P., Dolcourt, J. L. and Grant, L. D. (1980). Contribution of the care-giving environment to increased lead burden of children. *Am. J. Ment. Defic.*, **84**, 339

Milar, K. S., Krigman, M. R. and Grant, L. D. (1981). Effects of neonatal lead exposure on memory in rats. *Neurobehav. Toxicol. Teratol.*, **3**, 369

Miller, C. D., Buck, W. B., Hembrough, F. B. and Cunningham, W. L. (1982). Foetal rat development as influenced by maternal lead exposure. *Vet. Hum. Toxicol.*, **24**, 163

Minsker, D. H., Moskalski, N., Peter, C. P., Robertson, R. T. and Bokelman, D. L. (1979). Effects of lead exposure *in utero* or postpartum on brain histomorphology and behavior in rat offspring. *Teratology*, **19**, 40A

Minsker, D. H., Moskalski, N., Peter, C. P., Robertson, R. T. and Bokelman, D. L. (1982). Exposure of rats to lead nitrate *in utero* or postpartum; Effects on morphology and behavior. *Biol. Neonate*, **41**, 193

Mitchell, W. G., Cookson, S. L. and Mann, J. D. (1982). Effects of lead on behavoural responses to morphine in juvenile rats. Abstract from the *Annual Meeting of the Southern Society for Pediatric Research*, Jan 15–18, New Orleans. *Clin. Res.*, **29**, 892A

Modak, A. T., Purdy, R. H. and Stavinhoa, W. B. (1978). Changes in acetylcholine concentration in mouse brain following ingestion of lead acetate in drinking water. *Drug Chem. Toxicol.*, **1**, 373

Modak, A. T., Weintraub, S. T. and Stavinhoa, W. B. (1975). Effect of chronic ingestion of lead on the central cholinergic system in rat brain regions. *Toxicol. Appl. Pharmacol.*, **34**, 340

Momcilovic, B. and Kostial, K. (1974). Kinetics of lead retention and distribution in suckling and adult rats. *Env. Res.*, **8**, 214

von Monakow, C. (1880). Zur pathologischen anatomie der bleilahmung und der saturninen encephalopathie. *Arch. Psychiatr. Nervenkrie*, **10**, 495

Moncrieff, A. A., Clayton, B. E. and Roberts, E. E. (1967). Lead absorption in children. *Br. Med. J.*, **3**, 174

Moncrieff, A. A., Koumides, O. P., Clayton, B. E., Patrick, A. D., Renwick, A. G. C. and Roberts, G. E. (1964). Lead poisoning in children. *Arch. Dis. Child.*, **39**, 1

Moore, M. R. (1975). Lead and the mitochondrion. *Postgrad. Med. J.*, **51**, 760 and 774

Moore, M. R., Meredith, P. A. and Goldberg, A. (1977). A retrospective analysis of blood lead in mentally retarded children. *Lancet,* **1,** 717

Moore, M. R., Meredith, P. A. and Goldberg, A. (1980). Lead and heme biosynthesis. In Singhal, R. L. and Thomas, J. A. (eds.). *Lead Toxicity.* pp. 79–117. (Baltimore: Urban and Schwarzenberg)

Morse, D. L., Landrigan, P. J., Rosenblum, B. F., Hubert, J. S. and Honsworth, J. (1979). El Paso revisited. *J. Am. Med. Assoc.,* **242,** 739

Mullenix, P. (1977). Altered behavioral patterning in rats postnatally exposed to lead: The use of time-lapse photographic analysis. In Zenick, H. and Reiter, L. (eds.). *Behavioral Toxicity. An Emerging Discipline.* pp. (7) 1–13. (North Carolina: Government Printing House)

Mullenix, P. (1980). Effect of lead on spontaneous behavior. In Needleman, H. L. (ed.). *Low Level Lead Exposure: The Clinical Implications of Current Research.* (New York: Raven Press)

Murozami, M., Chow, T. J. and Patterson, C. C. (1969). Chemical concentrations of pollutant lead aerosols, terrestrial dusts, and sea salt in Greenland and Antarctic snow strata. *Geochim. Cosmochim. Acta,* **33,** 1247

Mykkanen, H. M., Lancaster, M. C. and Dickerson, J. W. T. (1982). Concentrations of lead in the soft tissues of male rats during a long-term dietary exposure. *Env. Res.,* **28,** 147

Nakagawa, K., Asami, M. and Kuriyama, K. (1980). Inhibition of release of lysosomal enzymes in young rat brain by lead acetate. *Toxicol. Appl. Pharmacol.,* **56,** 86

Nathanson, J. A. (1979). Chronic lead exposure, hyperactivity, and alteration of basal and catecholaminergic-sensitive adenylate cyclase. In *Catecholamines: Basic Clinical Frontiers. Proceedings of the Fourth International Catecholamine Symposium.* vol 2. pp. 1655–7 (New York: Pergamon)

Nathanson, J. A. and Bloom, F. E. (1975). Lead-induced inhibition of brain adenyl cyclase. *Nature (London),* **255,** 419

Needleman, H. L. (ed.). (1980). *Low Level Lead Exposure: the Clinical Implications of Current Research.* (New York: Raven Press)

Needleman, H. L., Gunnoe, C., Leviton, A., Reed, R., Peresie, H., Maher, C. and Barrett, P. (1979). Deficits in psychologic and classroom performance of children with elevated dentine lead levels. *N. Engl. J. Med.,* **300,** 689

Neuchay, B. R. and Saunders, J. P. (1978). Inhibitory characteristics of lead chloride in sodium and potassium ATP preparations derived from kidney, brain and heart of several species. *Toxicol. Env. Health,* **4,** 147

Nicander, (2nd Century BC). Theriaca and Alexipharmaca. (Translated into Latin by Euricius Cordus, Frankfurt, 1532). *See* Major, R. H. (1965)

Ohnishi, A., Schilling, K., Brimijoin, W. S., Lambert, E. H., Fairbanks, V. G. and Dyck, P. J. (1977). *J. Neuropathol. Exp. Neurol.,* **36,** 499

Okazaki, H., Aronson, S. M., DiMaio, D. J. and Olvera, J. E. (1963). Acute lead encephalopathy of childhood. Histological and chemical studies with particular reference to angiopathic aspects. *Trans. Am. Neurol. Assoc.,* **88,** 248

Oliver, T. (1891). *Lead Poisoning.* (Edinburgh: Bentland)

Oliver, T. (1914). *Lead Poisoning.* pp. 180–183. (London: H. K. Lewis)

Osheroff, M. R., Uno, H. and Bowman, R. E. (1982). Lead inclusion bodies in the anterior horn cells and neurons of the substanta nigra in the adult rhesus monkey. *Toxicol. Appl. Pharmacol.*, **64**, 570

Otto, D. A., Benignus, V. A., Muller, K. E. and Barton, C. N. (1981). Effects of age and body lead burdens on CNS function in young children. I. Slow potentials. *Electroencephalogr. Clin. Neurophysiol.*, **52**, 229

Overmann, S. R. (1977). Behavioral effects of asymptomatic lead exposure during neonatal development in rats. *Toxicol. Appl. Pharmacol.*, **41**, 459

Overmann, S. R., Zimmer, L. and Woolley, D. E. (1981). Motor development, tissue weights and seizure susceptibility in perinatally exposed rats. *Neurotoxicology*, **2**, 725

Padich, R. and Zenick, H. (1977). The effects of developmental and/or direct lead exposure on FR behaviour in the rat. *Pharmacol. Biochem. Behav.*, **6**, 371

Partington, J. R. (1934). *Origins and Development of Applied Chemistry*. (London: Longmans, Green and Co.)

Patel, A. J., Michaelson, I. A., Cremer, J. E. and Balazs, R. (1974a). The metabolism of [^{14}C]glucose by the brains of suckling rats intoxicated with inorganic lead. *J. Neurochem.*, **22**, 581

Patel, A. J., Michaelson, I. A., Cremer, J. E. and Balazs, R. (1974b). Changes within metabolic compartments in the brains of young rats ingesting lead. *J. Neurochem.*, **22**, 591

Patterson, C. C. (1980). An alternative perspective. In *Lead in the Human Environment*. p. 337. (Washington: US N.A.S.)

Patterson, C. C. (1983). British mega exposures to industrial lead. In Rutter, M. and Russell Jones, R. *Lead versus Health*. (London: Wiley)

Paul of Aegina (AD 625–690). *See* Adams, F. (1844–1847)

Paul, C. (1860). Etude sur l'intoxicatione lente par les preparations de plomb, de son influence sur produit de la conception. *Arch. Gen. Med.*, **15**, 513

Pentschew, A. (1965). Morphology and morphogenesis of lead encephalopathy. *Acta Neuropathol. (Berlin)*, **5**, 133

Pentschew, A. and Garro, F. (1966). Lead encephalo-myelopathy of the suckling rat and its implications on the porphrynopathic nervous diseases. *Acta Neuropathol. (Berlin)*, **6**, 266

Perlstein, M. A. and Attala, R. Neurologic sequelae of plumbism in children. *Clin. Pediatr.*, **5**, 292

Petit, T. L. and Alfano, D. P. (1979). Differential experience following developmental lead exposure: effects on brain and behavior. *Pharmacol. Biochem. Behav.*, **11**, 165

Petit, T. L. and LeBoutillier, J. C. (1979). Effects of lead exposure during development on neocortical dendritic and synaptic structure. *Exp. Neurol.*, **64**, 482

Petrusz, P., Weaver, C. M., Grant, L. D., Mushak, P. and Krigman, M. R. (1979). Lead poisoning and reproduction: Effects on pituitary and serum gonadotropins in neonatal rats. *Env. Res.*, **19**, 383

Peuschel, S. M., Kopito, L. and Schwachmann, H. (1972). Children with an increased lead burden. *J. Am. Med. Assoc.*, **222**, 462

Piepho, R. W., Ryan, C. F. and Lacz, J. P. (1976). The effects of chronic lead intoxication on the gamma-aminobutyric acid content of the rat CNS. *Pharmacologist*, **18**, 125 (abstract 062)

Pihl, R. O. and Parks, M. (1977). Hair element content in learning disabled children. *Science*, **198**, 204

Piomelli, S. (1980). Effects of low level lead exposure on heme metabolism. In Needleman, H. L. (ed.). *Low Level Lead Exposure: The Clinical Implications of Current Research.* pp. 67–74. (New York: Raven Press)

Pliny (the Elder). *See* Jones, W. H. S. and Rackham, H. (1938–1963)

Press, M. (1977a). Animal model of human disease: Lead encephalopathy in neonatal Long–Evans rats poisoned via an eosophageal catheter, *Am. J. Pathol.*, **86**, 485

Press, M. (1977b). Lead encephalopathy in neonatal Long–Evans rats. Morphologic studies. *J. Neuropathol. Exp. Neurol.*, **36**, 169

Press, M. (1977c). Neuronal development in the cerebellum of lead-poisoned neonatal rats. *Acta Neuropathol. (Berlin)*, **40**, 259

Price, (1978). *Price's Textbook of the Practice of Medicine.* 12th Edn. Scott, Sir R. B. (ed.). (Oxford: Oxford University Press)

Purdy, S. E., Blair, J. A., Leeming, R. J. and Hilburn, M. E. (1981). Effect of lead on tetrahydrobiopterin synthesis and salvage: a cause of neurological dysfunction. *Int. J. Env. Stud.*, **17**, 141

Rafales, L. S., Greenland, R. D., Zenick, H., Goldsmith, M. and Michaelson, I. A. (1981). Responsiveness to D-amphetamine in lead-exposed rats as measured by steady state levels of catecholamines and locomotor activity. *Neurobehav. Toxicol. Teratol.*, **3**, 363

Ramsay, P. B., Krigman, M. R. and Morrell, P. (1980). Developmental studies of the uptake of choline, gaba and dopamine by crude synapsomal preparations after *in vivo* or *in vitro* lead treatment. *Brain Res.*, **187**, 383

Ratcliffe, J. M. (1977). Developmental and behavioural functions in young children with elevated blood lead levels. *Br. J. Preventive Social Med.*, **31**, 258

Reiter, L. W. (1977). Behavioral toxicology: effects of early postnatal exposure to neurotoxins on development of locomotor activity in the rat. *J. Occupational Med.*, **19**, 201

Reiter, L. W., Anderson, G. E., Laskey, J. W. and Cahill, D. F. (1975). Developmental and behavioral changes in the rat during chronic exposure to lead. *Env. Health Perspectives*, **12**, 119

Reiter, L. W. and Ash, M. E. (1976). Neurotoxicity during lead exposure in the rat. *Toxicol. Appl. Pharmacol.* **37**, 160 (abstract 163)

Reyners, H., de Reyners, E. G. and Maisin, J.-R. (1976). Hypervascularization of the cerebral cortex in lead-induced encephalopathy. *Experientia*, **32**, 1416

Reyners, H., de Reyners, E. G. and Maisin, J.-R. (1979). An ultrastructural study of the effects of lead in the central nervous system. *International Conference on the Management and Control of Heavy Metals in the Environment*, pp. 58–61

Reyners, H., de Reyners, E. G., Tachon, P., Laschi, A. and Maisin, J.-R. (1980). Lead encephalopathy in the adult monkey: an ultrastructural approach. *2nd International Congress on Toxicology*, July, 6–11, Brussels, *Toxicol. Lett. Special Issue*, **0**, 76 (abstract 0. 122)

Reyners, H., de Reyners, E. G., van der Pareren, J. and Maisin, J.-R. (1978). Evolution de l'equilibre des populations gliales dans le cortex cerebral du rat intoxique au plomb. *C. R. Soc. Belge Biol.*, **172**, 998

Rice, D. C. and Willes, R. F. (1979). Neonatal low level lead exposure in monkeys *(Macaca fascicularis)*: effect on two choice nonspatial form discrimination. *J. Env. Pathol. Toxicol.*, **2**, 1195

Roels, H. A., Buchet, J. P., Lauwerys, R., Bruax, P., Claeys-Thorsau, F., Lafontaine, A., van Overschelde, J. and Verduyn, G. (1978). Lead and cadmium absorption among children near a nonferrous metal plant. *Env. Res.,* **15,** 290.

Rosenblum, W. J. and Johnson, M. G. (1968). Neuropathologic changes produced in suckling mice by adding lead to the maternal diet. *Arch. Pathol.,* **85,** 640

Royal Commission on Environmental Pollution. (1983). *Ninth Report. Lead in the Environment.* (London: HMSO)

Rummo, J. H., Horn, D. K., Rummo, N. J. and Brown, J. F. (1979). Behavioral and neurological effects of symptomatic and asymptomatic lead exposure in children. *Arch. Env. Health,* **34,** 120

Rutter, M. (1980). Raised lead levels and impaired cognitive/behavioural functioning: A review of the evidence. *Dev. Med. Child Neurol.,* **22** (Supplement 42), 1

Sachs, H. K. (1974). Effect of a screening programme on changing patterns of lead poisoning. *Env. Health Perspectives,* **7,** 41

Sachs, H. K., Blanksma, L. A., Murray, E. F. and O'Connell, M. J. (1970). Ambulatory treatment of lead poisoning report of 1155 cases. *Pediatrics,* **46,** 389

Sachs, H. K., Krall, V. and Drayton, M. A. (1982). Neuropsychological assessment after lead poisoning without encephalopathy. *Perceptual Motor Skills,* **54,** 1283

Sachs, H. K., Krall, V., McGaughran, D. A., Rozenfeld, I. H., Yongsmith, N., Growe, G., Lazar, B. S., Novar, L., O'Connell, L. and Rayson, B. (1978). IQ following treatment of lead poisoning: A patient-sibling comparison. *J. Pediatr.,* **93,** 428

Santos-Anderson, R. M., Tso, M. O. M., Valdes, J. and Annua, Z. (1980). Retinal pathology in chronic lead poisoning. *Assoc. for Res. Vision Ophthalmol. Abstr. Invest. Ophthalmol. Visual Sci.,* **20,** (Supplement 3), 80 (abstract 40)

Sauerhoff, M. W. and Michaelson, I. A. (1973). Hyperactivity and brain catecholamines in lead-exposed developing rats. *Science,* **182,** 1022

Scarborough, J. (1969). *Roman Medicine.* (New York: Cornell University Press)

Schlaepfer, W. W. (1969). Experimental lead neuropathy: a disease of the supporting cells in the peripheral nervous system. *J. Neuropathol. Exp. Neurol.,* **21,** 401

Schroeder, H. A. and Mitchener, M. (1971). Toxic effects of trace elements on the reproduction of mice and rats. *Arch. Env. Health,* **23,** 102

Schroeder, H. A. and Tipton, I. H. (1968). The human body burden of lead. *Arch. Env. Health,* **17,** 965

Schumann, A. M. (1977). The effects of inorganic lead on the central catecholaminergic system of the rodent with emphasis on postnatally exposed rats and mice. *Diss. Abstr. Int.,* **38,** 5880-B

Schupe, J. L., Binns, W., James, L. F. and Keeler, R. F. (1967). Lupine, a case of crooked calf disease. *J. Am. Vet. Med. Assoc.,* **151,** 198

Seigel, G. J., Fogt, S. K. and Hurley, M. J. (1977). Lead actions on sodium plus potassium adenine triphosphatase from electroplax, rat brain and rat kidney. In Miller, M. W. and Shamoo, A. E. (eds.). *Membrane Toxicity.* pp. 465–93. *Adv. Exp. Med. Biol.,* **84** (New York: Plenum Press)

Selhi, H. S. and White, J. M. (1975). The effect of lead on the red cell membrane. *Postgrad. Med. J.,* **51,** 765

de Selincourt, A. (1966). *Heroditus: The Histories.* p. 88 (Harmondsworth: Penguin)

Seppalainen, A. M., Tola, S., Hernberg, S. and Kock, B. (1975). Subclinical neuropathy at 'safe' levels of lead exposure. *Arch. Env. Health,* **30,** 180

Sharding, N. N. and Oehme, F. W. (1973). The use of animal models for comparitive studies of lead poisoning. *Clin. Toxicol. Bull.*, **3**, 103

Shearer, T. R., Larson, K., Neuschwander, J. and Gedney, B. (1982). Minerals in the hair and nutrient intake of autistic children. *J. Autism Dev. Disord.*, **12**, 25

Shigeta, S., Misawa, T., Aikawa, H. and Yokoyama, M. (1980). Effects of learning schedules on operant behavior in lead-administered rats. *Jap. J. Hyg.*, **35**, 752

Shih, T.-M. and Hanin, I. (1977). Lead exposure decreases acetylcholine turnover rate in rat brain areas *in vivo. Fed. Proc.*, **36**, 977 (abstract 3733)

Shih, T.-M. and Hanin, I. (1978a). Effects of chronic lead exposure on levels of acetylcholine and choline and on acetylcholine turnover rate in rat brain areas *in vivo. Psychopharmacology*, **58**, 263

Shih, T.-M. and Hanin, I. (1978b). Chronic lead exposure in immature animals: neurochemical correlates. *Life Sci.*, **23**, 877

Shirabe, T. and Hirano, A. (1977). X-ray microanalytical studies of lead-implanted rat brains. *Acta Neuropathol. (Berlin)*, **40**, 189

Silbergeld, E. K. (1977). Interactions of lead and calcium on the synaptosome uptake of dopamine and choline. *Life Sci.*, **20**, 309

Silbergeld, E. K. and Adler, H. S. (1977). Subcellular mechanisms of lead neurotoxicity. *Brain Res.*, **148**, 451

Silbergeld, E. K., Adler, H. S. and Costa, J. L. (1977). Subcellular localization of lead in synaptosomes. *Res. Commun. Chem. Pathol. Pharmacol.*, **17**, 715

Silbergeld, E. K. and Chisholm, Jnr., J. J. (1976). Lead poisoning: altered urinary catecholamine metabolites as indicators of intoxication in mice and children. *Science*, **192**, 153

Silbergeld, E. K. and Goldberg, A. M. (1973). A lead-induced behavioral disorder. *Life Sci.*, **13**, 1275

Silbergeld, E. K. and Goldberg, A. M. (1974a). Lead-induced behavioral dysfunction: an animal model of hyperactivity. *Exp. Neurol.*, **42**, 146

Silbergeld, E. K. and Goldberg, A. M. (1974b). Hyperactivity: a lead-induced-behavior disorder. *Env. Health Perspectives*, **7**, 227

Silbergeld, E. K. and Goldberg, A. M. (1975). Pharmácological and neurochemical investigations of lead-induced hyperactivity. *Neuropharmacology*, **14**, 431

Silbergeld, E. K. and Hruska, R. E. (1980). Neurochemical investigations of low level lead exposure. In Needleman, H. L. (ed.). *Low Level Lead Exposure: The Clinical Implications of Current Research*. pp. 135–57. (New York: Raven Press)

Silbergeld, E. K., Hruska, R. E., Miller, L. P. and Eng, N. (1980). Effects of lead *in vivo* on GABAergic neurochemistry. *J. Neurochem.*, **34**, 1712

Silbergeld, E. K. and Lamon, J. M. (1980). The role of altered haem synthesis in the neurotoxicity of lead. *J. Occup. Med.*, **25**, 680

Silbergeld, E. K., Miller, L. P., Kennedy, S. and Eng, N. (1979). Lead, GABA, and seizures: Effects of subencephalopathic lead exposure on seizure sensitivity and GABAergic function. *Env. Res.*, **19**, 371

Simpson, J. A., Seaton, D. A. and Adams, J. F. (1964). Response to treatment with chelating agents of anaemia, chronic encephalopathy, and myelopathy due to lead poisoning. *J. Neurol. Neurosurg. Psychiatr.*, **27**, 536

Singhal, R. L. and Thomas, J. A. (eds.). (1980). *Lead Toxicity*. (Baltimore: Urban and Schwarzenberg)

Smith, J. F., McLaurin, R. L., Nichols, J. B. and Asbury, A. (1960). Studies in cerebral edema and cerebral swelling. I. The changes in lead encephalopathy in children compared with those in alkyltin poisoning in animals. *Brain,* **83,** 411

Snowdon, C. T. (1973). Learning deficits in lead-injected rats. *Pharmacol. Biochem. Behav.,* **1,** 599

Sobotka, T. J., Brodie, R. E. and Cook, M. P. (1975). Psychophysiologic effects of early lead exposure. *Toxicology,* **5,** 175

Sobotka, T. J. and Cook, M. P. (1974). Postnatal lead acetate exposure in rats: possible relationship to minimal brain dysfunction. *Am. J. Mental Deficiency,* **79,** 5

Srivastava, U. and Thakur, M. L. (1981). Effect of lead exposure on the cellular growth of the brain in the rat progeny. In Tsang, R. C. and Nichols, B. L. (eds.). *Nutrition and Child Health: Perspectives for the 1980s.* p. 191. (New York: Liss)

Stephens, M. C. C. and Gerber, G. B. (1981). Development of glycolipids and gangliosides in lead-treated neonatal rats. *Toxicol. Lett.,* **7,** 373

Sterling, G. H., O'Neill, K. J., McCafferty, M. R. and O'Neill, J. J. (1982). Effect of chronic lead ingestion by rats on glucose metabolism and acetylcholine synthesis in cerebral cortex slices. *J. Neurochem.,* **39,** 592

Stowe, H. D. and Vandevelde, M. (1974). Lead-induced encephalopathy in dogs fed high fat, low calcium diets. *J. Neuropathol. Exp. Neurol.,* **38,** 463

Stumpf, W. E., Sar, M. and Grant, L. D. (1980). Autoradiographic localization of ^{210}Pb and its decay products in rat forebrain. *Neurotoxicology,* **1,** 593

Takeichi, M. and Nada, Y. (1974). Electron microscopy of experimental lead encephalopathy. Consideration on the development mechanism of brain lesions. *Folia Psychiatr. Neuro. Jap.,* **28,** 217

Tanquerel des Planches, L. *See* Dana, S. L. (1848)

Taylor, D., Nathanson, J., Hoffer, B., Olson, L. and Seiger, A. (1978). Lead blockade of norepinephrine-induced inhibition of cerebellar Purkinje neurons. *J. Pharmacol. Exp. Ther.,* **206,** 371

Taylor, D. H., Noland, E. A., Brubaker, C. M., Crofton, K. M. and Bull, R. J. (1982). Low level lead (Pb) exposure produces learning deficits in young rat pups. *Neurobehav. Toxicol. Teratol.,* **4,** 311

Tennekoon, G., Aitchinson, C. S., Frangia, J., Price, D. L. and Goldberg, A. M. (1979). Chronic lead intoxication: effects on developing optic nerve. *Ann. Neurol.,* **5,** 558

Tesh, J. M. and Pritchard, A. L. (1980). Lead – a behavioural teratogen. *Acta Morphol. Acad. Sci. Hung.,* **28,** 227

Thomas, H. F., Elwood, P. C., Welsby, E. and St. Leger, A. S. (1979). Relationship of blood lead in women and children to domestic water lead. *Nature (London),* **282,** 712

Thomas, J. A., Dallenbach, F. D. and Thomas, M. (1971). Considerations on the development of experimental lead encephalopathy. *Virchows Arch. Abteilung A. Pathol. Anat.,* **352,** 61

Thomas, J. A., Dallenbach, F. D. and Thomas, M. (1973). The distribution of radioactive lead [^{219}Pb] in the cerebellum of developing rats. *J. Pathol.,* **109,** 45

Thomas, J. A. and Thomas, M. (1974). The pathogenesis of lead encephalopathy. *Ind. J. Med. Res.,* **62,** 36

Toews, A. D., Kolber, A., Hayward, J., Krigman, M. R. and Morrell, P. (1978). Experimental lead encephalopathy in the suckling rat: concentration of lead in cellular fractions enriched in brain capillaries. *Brain Res.,* **147,** 131

Toews, A. D., Krigman, M. R., Thomas, D. J. and Morrell. P. (1980). Effect of inorganic lead exposure on myelination in the rat. *Neurochem. Res.*, **5**, 605

Tredgold, A. F. (1947. *Mental Deficiency* 7th Edn. (London: Ballieve, Tyndall and Cox)

Tronchin, T. (1709–1781). De Colica Pictonum. Geneva: Cramer (1757). *See* Major, R. H. (1965)

UK D.H.S.S. (1980). *Lead and Health: The report of a DHSS working party on Lead in the environment* (London: HMSO)

UK D.o.E. (1974). *Lead Pollution in Birmingham.* Department of the Environment (London: HMSO)

US C.D.C. (1978). *Preventing Lead Poisoning in Children.* (Atlanta: Center for Disease Control, US Department of Health, Education and Welfare)

US E.P.A. (1977). *Air Quality Criteria for Lead.* (Washington: US Environmental Protection Agency)

US N.A.S. (1972). *Lead: Airborne Lead in Perspective.* (Washington: National Academy of Sciences)

US N.A.S. (1978). *Lead in the Human Environment.* (Washington: National Academy of Sciences)

Vanderkooi, J. M. and Landesberg, R. *In vivo* synthesis of iron-free cytochrome C during lead intoxication. *FEBS Lett.*, **73**, 254

Verlangieri, A. J. (1979). Prenatal and postnatal chronic lead intoxication and running wheel activity in the rat. *Pharmacol. Biochem. Behav.*, **11**, 95

Verschaeve, L., Driesen, M., Kirsch-Volders, M., Hans, L. and Susanne, C. (1979). Chromosome distribution studies after inorganic lead exposure. *Hum. Genet.*, **49**, 147

Vistica, D. T. and Ahrens, F. A. (1977). Microvascular effects of lead in the neonatal rat. II. An ultrastructural study. *Exp. Mol. Pathol.*, **26**, 139

Waldron, H. A. (1973). Lead poisoning in the ancient world. *Med. Hist.*, **17**, 391

Waldron, H. A. and Stofen, D. (1974). *Sub-Clinical Lead Poisoning.* (London and New York: Academic Press)

Walsh, R. N. and Cummins, R. A. (1976). The open-field test: a critical review. *Psychol. Bull.*, **83**, 482

Walter, S. D., Yankel, A. J. and von Lindern, I. H. (1980). Age-specific risk factors for lead absorption in children. *Arch. Env. Health,* **35**, 53

Warner, R. (1961). *Xenephon: The Persian Expedition.* (Harmondsworth: Penguin)

W.B.M.S. (1980). *World Metal Statistics.* (London: World Bureau of Metal Statistics)

Weinreich, K., Stelte, W. and Bitsch, I. (1977). Effect of lead on the spontaneous activity of young rats. *Second European Nutrition Conference,* Munich, 1976. *Nutr. Metabol.*, **21**, (Supplement 1), 201

Wells, G. A. H., McHowell, J. and Gopinath, C. (1976). Experimental lead encephalopathycalves. Histological observations on the nature and distribution of lesions. *Neuropathol. Appl. Neurobiol.*, **2**, 605

Wender, P. H. (1978). Minimal brain dysfunction: an overview. In Lipton, M. A., DiMascio, A. and Killam, K. F. (eds.). *Psychopharmacology: A Generation of Progress.* (New York: Raven Press)

Werboff, J. and Gottlieb, J. S. (1963). Drugs in pregnancy: behavioural teratology. *Obstet. Gynaecol. Survey,* **18**, 420

Whitfield, C. L., Chi'en, L. T. and Whitehead, J. D. (1972). Lead encephalopathy in adults. *Am. J. Med.,* **52,** 289

W.H.O. (1977). *Environmental Health Criteria. 3. Lead* (Geneva: World Health Organization)

Wickham, E. C. (1891). *The Works of Horace, with a Commentary.* (Oxford: Clarendon Press)

Wiener, G. (1970). Varying psychological sequelae of lead ingestion in children. *Public Health Rep.,* **85,** 19

Wilson, S. A. K. (1954). In Bruce, A. N. *Neurology.* 2nd Edn. p. 850–1. (London: Butterworths)

Wince, L. C. and Azzaro, A. J. (1977). Dopamine synaptic function following low-level lead carbonate exposure in neonatal rats. *Pharmacologist,* **19,** 134 (abstract 041)

Wince, L. and Azzarro, A. J. (1978). Neurochemical changes of the central dopamine synapse following chronic lead exposure. *Neurology,* **28,** 382 (abstract PP46)

Wince, L. C., Donovan, C. H. and Azzaro, A. J. (1976). Behavioral and biochemical analysis of the lead-exposed hyperactive rat. *Pharmacologist,* **18,** 198 (abstract 473)

Wince, L. C., Donovan, C. A. and Azzaro, A. J. (1980). Alterations in the biochemical properties of central dopamine synapses following chronic postnatal PbCO$_3$ exposure. *J. Pharmacol. Exp. Ther.,* **214,** 642

Winder, C., Garten, L. L., Jordan, T. C. and Lewis, P. D. (1984a). Hippocampal function in adult rats perinatally exposed to lead (Submitted to *Pharmacol. Biochem. Behav.)*

Winder, C., Kitchen, I., Clayton, L. B., Gardener, S., Wilson, J. and Lewis, P. D. (1984b). The effect of perinatal lead administration on the ontogeny of striatal enkephalin levels in the rat. (Submitted to *Toxicol. Appl. Pharmacol.)*

Winder, C., Hollins, G. W., Lewis, P. D. and Garten, L. L. (1984c). Effects of perinatal low level lead exposure on the rat hippocampus. A paper presented at the *British Neuropathological Society,* July 7–8, Glasgow, Scotland (Neuropathol. Appl. Neurobiol. In press)

Winder, C., Carmichael, N. G., Garten, L. L. and Lewis, P. D. (1984d). Dose response relationships during perinatal lead administration in the rat: a reassessment and redefinition. (In preparation)

Winneke, G. (1981). Neuropsychologische bleiwirkungen bei kindern. Eine ubersicht. In Globel, G. (ed.). Das Strahlenrisiko im Vergleich zu Chemischen und Biologischen Risiken. p. 356–67. (Stuttgart: Thieme)

Winneke, G., Brockhaus, A. and Baltissen, R. (1977). Neurobehavioral and systemic effects of longterm blood lead-elevation in rats. *Arch. Toxicol.,* **37,** 247

Winneke, G., Lilienthal, H. and Werner, W. (1982). Task-dependent neurobehavioral effects of lead in rats. New Toxicology for Old. *Arch. Toxicol. Suppl.,* **5,** 84

Xenephon (400 BC) *See* Warner (1961)

Yule, W., Lansdown, R., Millar, I. B. and Urbanowicz, M.-A. (1981). The relationship between blood lead concentrations, intelligence and attainment in a school population: pilot study. *Dev. Med. Child Neurol.,* **23,** 567

Zenick, H. and Goldsmith, M. (1981). Drug discrimination learning in lead-exposed rats. *Sci.,* **212,** 569

Zenick, H., Lasley, S. M., Greenland, R., Caruso, V., Succop, P., Price, D. and Michaelson, I. A. (1982). Regional brain distribution of D-amphetamine in lead-exposed rats. *Toxicol. Appl. Pharmacol.*, **64**, 52

Zenick, H., Lasley, S., Greenland, R. and Michaelson, I. A. (1981). Brain levels and attenuated response to D-amphetamine in lead exposed rats. *Fed. Proc.*, **40**, 700

Zenick, H., Padich, R., Tokarek, T. and Aragon, P. (1978). Influence of prenatal and postnatal lead exposure on discrimination learning in rats. *Pharmacol. Biochem. Behav.*, **8**, 346

Zenick, H., Pecorraro, F., Price, D., Saez, K. and Ward, J. (1979a). Maternal behaviour during chronic lead exposure and measures of offspring development. *Neurobehav. Toxicol.*, **1**, 65

Zenick, H., Rodriquez, W., Ward, J. and Elkington, B. (1979b). Deficits in fixed interval performance following prenatal and postnatal lead exposure. *Dev. Psychobiol.*, **12**, 509

Zenick, H., Ward, J., Rodriguez, W., Aragon, P. and Scrivseth, R. (1979c). Offspring open field performance following maternal lead exposure: a question of dosage and nutritional status. *Pharmacol. Biochem. Behav.*, **11**, Suppl., 35

Zielhuis, R. L. (1975). Dose-response relationships for inorganic lead. *Int. Arch. Occup. Health*, **35**, 1

Zimmering, R. T., Burright, R. G. and Donovick, P. J. (1982). Effects of prenatal and continued lead exposure on activity levels in the mouse. *Neurobehav. Toxicol. Teratol.*, **4**, 9

Index

abortifacient 22
abortion, lead workers 22
acetylcholine 83, 84, 93
acetylcholinesterase 83, 85, 93
acetyl coenzyme A 93
acquisition 125
  definition 126
activity cage
  Animex 56
  automated 62
  plexiglass 56
activity meter 55, 56
adenylate cyclase
  dopamine receptor 88, 96, 99
  lead inhibition 82
  metabolism 88
  postsynaptic 94
  receptor linked 88, 96
adrenergic responses 88
age
  blood levels and lead administration
    46–8
  gut lead absorption 67
alkaline phosphatase 78
alloys 3
amino acids, brain levels and lead 99
γ-aminobutyric acid (GABA) 83
  cerebellar levels 89, 97
  inhibitory neurotransmitter 98
  potassium-stimulated uptake 90, 97
  system, effects of lead 71, 89–91, 97, 98
γ-aminobutyric acid transaminase (GABA-
    transaminase) lead inhibition 91,
    97, 98
D-aminolaevulinic acid
  GABA interactions 98, 127
  excretion 12, 63
D-aminolaevulinic acid dehydratase, lead
    inhibition 78, 81
amniotic fluid, lead levels 23
AMP, cyclic 99
amphetamine
  -induced locomotor activity and lead 69

lead effects on behaviour 59, 60, 64, 70
  uptake 87
amulets 8
amygdala, lead accumulation 49
anglesite 2
animal models
  administered dose 45
  applicability 37
  dose and growth 45
  dose of lead 38–42
  extrapolation 124, 125
  lead encephalopathy 35
  lead intoxication 34–45
  lead sources 38–42
  Pentschew and Garro 35, 36, 104
  route and duration of lead 38–42
  susceptibility 123
  toxic responses, factors affecting 45, 46
  usefulness 123
  weaned diet 36
antiknock agents 4
anus, blue line 12
apomorphine and lead effects on
    behaviour 60, 69
arousal levels 65, 73
aspartate, brain 77, 99
astrocytes 106
  lead inclusions 51
  PAS-positive granules 113
  proliferation 19, 110, 115
astroglia 119
ATPase 82
  lead site of action 82
atropine
  action 71
  drug effects on behaviour 61, 71
audiogenic seizure 59
auditory startle response 68
autism 29
avoidance tasks 58

basement membrane changes 110, 114
battery factory, child lead effects 29, 30

behaviour *see also* individual parameters,
    tests, hyperactivity
  activity test systems 53–6, 62–4, 126
  aggressive 63
  agonistic 59
  alternation tasks 64
  CNS function 125
  complex 59, 68, 73
  conditioned 64
  drug-induced 59–61, 68–72, 100
  effects of lead 53–73
  exploratory activity, lead timing 64
  maternal 59
  simple 53–68
  social 59
  task selection 72
  teratology 22
  water lead levels 66
benztropine
  action 71
  lead effects on behaviour 61, 69, 71
Betts process 3
Bible, cupellation references 7, 8
biopterin synthesis
  lead effects 81, 82, 100
  salvage pathway 82
blood-brain barrier 104
blood levels
  abnormal
    criticisms 34
    EEC 33
    US 33
  brain accumulation 49
  clinical limit 1
  elevated lead 24
  encephalopathy 13, 20, 120, 121
  energy metabolism and lead 93
  environmental risks 29, 30
  exposure index 46
  factors affecting 45, 46
  indicator 27
  lead dose, duration and age 46–8
  lead poisoning 24
  measurement 46
  mental retardation 28
  neuropathological findings 121
  neuropsychological tests 30
  normal values 1, 27, 33
  prenatal 65
  psychometric tests 26, 28
  recommendations 33
  steady state 46–8
  urban children 23, 24
blood–nerve barrier breakdown 102
boutons 107
brain
  biochemical parameters
    investigated 76–9

glucose and acetate metabolism 77, 80
lead distribution 49–51
lead uptake and regions 50
mitochondria 80
morphological effects of lead 105–12 *see
    also* brain regions
oedema 109, 114
  source 119
offspring and parental lead 62, 63
protein and lead 75, 76, 80
RNA and DNA 76, 77
stem, lead levels 49, 50
weight and lead 105
brainstem changes and lead 112
Burtonian line 12
butyrylcholinesterase, brain activity and
  lead 85, 93

calcium, competitive inhibition 80, 93
  dopamine release 94
capillary development
  lead 110, 114
  lead concentration 114
capillary endothelium damage 113
capillary permeability 104, 106, 110
catalase 78
catecholaminergic systems
  effects of lead 86–9, 94–7
    types 97
  lead effects on behaviour 60, 61, 69, 70
  turnover assessments 95
cathepsin 78
cellular distribution of lead, 51
cerebellar haemorrhage 37, 49, 109, 110
cerebellum
  later development 117
  lead concentration and effects 50,
    108–12, 116, 117
  lead damage 75, 103, 108, 109, 114, 115
  low dose lead 119
cerebral cortex
  capillaries 106, 110, 114
  dendrite alterations 106, 107, 118
  lead damage 75, 103–7, 114, 117
  lead uptake 50, 51
  vascular damage 104–6, 114
cerebrosides and lead 79
ceruse uses and toxicity 9, 10
cerussite 2
chelating agents and hyperactivity 28
chloral hydrate sedation and lead 61, 71
chlorpromazine and sedation with lead 61,
  71
cholesterol 79
choline acetyltransferase, brain activity and
  lead 83, 85, 93
choline binding, levels and lead 83, 84

choline phosphokinase, brain levels and
    lead 83, 86, 93
cholinergic agents, lead effects 71
cholinergic neurotransmitters
    lead effects 81, 83–6, 93
    metabolism 83, 93
    pathological changes 93
cholinesterase 85
chromosomes, lead effects 23
CNS
    function 20, 21, 125
    high lead studies 104–19
    lead distribution 49
    lead uptake 50
    low lead studies 105, 107, 108, 119, 120
    morphological lead effects 101
        human autopsy 101
cocaine 60
cognitive function 73
colic 9, 12, 13
    epidemics 22
colostrum 36
complexing groups, lead 81
continuous reinforcement 58
corpus callosum changes and lead 108
corpus striatum changes and lead 108
cross-fostering 64
cupellation 7
    origin 5
cytochrome metabolism, brain and
    lead 78, 80, 81

demyelination, segmental 103
dentate fascia 49, 117, 120
dentate gyrus 117
dentine lead 26, 33
Denver Developmental Screening Test 25
development and lead effects 34
Devonshire colic and cider 11
diet
    lead administration and animals 36
    lead uptake 17, 18
    palatability and lead 45 see also
        undernutrition
differential reinforcement 58, 67, 68, 73
dihydrobiopterin 82
dihydroxyphenylacetic acid (DOPAC)
    brain levels and lead 87, 96
dimethylaminoethanol and lead effects on
    behaviour 61, 71
discrimination tasks 56, 63, 65, 66, 73
distillation apparatus and lead 11
DNA, brain and lead 75, 76, 80
L-DOPA, lead effects on behaviour 60, 69
    metabolism 82
DOPAC, brain levels 87, 96
dopamine
    metabolism and lead 86, 94

metabolites 87, 95
    synaptosome release 94
    transport and lead 95
dopamine-β-hydroxylase 97
drug-induced behaviour and lead
    catecholaminergic effects 59, 60, 69, 70
    indirect effects 70
    cholinergic agent effects 61, 71
    GABAergic agents 71
Dutch process 3

EA 32138, oldest lead artefact 5, 6
electroencephalogram (EEG) and lead
    exposure 30
electroshock seizure test 59, 68
E-maze 57, 63
emotionality
    assessment 62
    lead dose 54, 62
encephalopathy 34, 124
    acute, features 115
    animal models 35, 103
    blood levels 121
        childhood 20, 121
        rat 120
    brain effects 19, 101
    brain structural changes 20
    features in adult animals 35
    high dose lead 104, 128
    mental impairment 25, 26
    onset and blood levels 20
    saturnine and blood levels 13
endoplasmic reticulum, vacuolation 113
energy metabolism
    brain and lead 77, 78, 80
    uncoupling 81
enkephalins, endogenous brain levels and
    lead 92, 98
enriched environments 56
environmental exposure and hazards 29–31
enzyme activity 9
    lead sensitivity 81
epidemiology, environmental hazards
    29–31
erythrocytes and CNS damage 49
excretion 18, 19
exhaust emission 4
exploratory activity 64
exposure, definition 33
    blood lead 27
    intrauterine 29
    paediatric, critical periods 37
extraction methods 2, 3
eye opening 54, 68
eye paint 8

Factory and Workshop Act (1895) 13
Fanconi syndrome 13

157

fenfluramine, lead effects on
    behaviour 60, 69
ferrochelatase 81
fetal toxicity 23
fixed interval schedule 58, 67
fixed ratio responding 66, 67
flinch-jump test 59
forearm muscles 101, 102
forebrain
  lead levels 50
  myelin 115
    weight, DNA and lead 105

GABA *see* γ-aminobutyric acid
galena 2, 5, 8
gangliosides 79
gavage 36, 114, 115
  lead administration 63
glial cell effects
  lead 106, 110, 111, 115, 116
  proliferation 104, 115
gliogenesis 119
glutamic acid, brain 76, 77, 99
  decarboxylase inhibition 91, 98
glutamine, brain 99
glutathione, brain and lead 78
glycine, brain 76, 99
glycolytic enzymes, brain and lead 77, 80
grooming, lead responses 55, 62
growth retardation and brain damage 114
guanylate cyclase 91, 96

haemorrhage, brain and lead 19, 35, 101,
    105, 109, 110, 114
  sodium and potassium changes 82
hair, lead exposure determination 29, 33
haloperidol and lead effects on
    behaviour 60, 69
hanging gardens of Babylon 7
Hebbs-Williams closed field maze 57, 66,
    126
high-dose lead studies, CNS effects 104–19
hilus 49
hind limb paralysis 103
hippocampus
  behavioural effects mediation 126
  dentate granule cells 118
  lead accumulation 50, 107, 117, 118
  lead distribution 49–51
    dentate hilus 117
    lead-related changes 107, 118
  lead effects 65, 118
  lead vulnerability 117
  low dose lead 119
  mossy fibre development 107, 118
Hippocrates, lead poisoning 9
histidine, brain 76
historical uses 5–14

home cage exploration apparatus 56
home environment 31
homovanillic acid (HVA)
  lead exposure 95
  lead and urine levels 87, 95
  mice and childrens response 21, 96
house dust 30
5-hydroxyindoleacetic acid (HIAA)
  endogenous levels and lead 89
  lead workers urine 21
5-hydroxytryptamine (5-HT, serotonin)
  endogenous levels and lead 99
  system 83
hyperactivity 28, 30, 53, 62, 99
  assessment methods 54–6
  lead dose in animals 53–6
  model 124
  morphine 72
hypomyelination 115
hypothalamus 50

ignition process, petrol engine 4
inclusion bodies 51
infrared photodetector array 56
intestinal absorption 46
intranuclear inclusion bodies 51
intrauterine exposure 29
IQ
  behaviour 27
  blood lead exposure 26–8, 30
ischaemia 19
isoniazid and convulsions with lead 61, 71

jiggle cage 55

Lashley jumping stand 57
lead carbonate 2
  uses 3
lead chromate, uses 3
lead colic, synonyms 11
lead intoxication
  acute, clinical features 12
  chronic, clinical features 12, 13
  screening programmes 24
  suckling rats 35
lead monoxide (litharge) uses 3
lead and opium solution 11
lead poisoning
  deaths 24
early descritions 9, 10
  first clinical descriptions 10
  occupational 13
  subclinical
    blood levels 45
    concept development 33-51
    definition 34
  subsequent neurological damage 25

lead toxicity
  paternal and fetus 22
  subclinical effects 24
lead workers
  miscarriages 22
  nerve damage 102
learning
  appetitive 56
  lead effects on animal 56–9, 64–8
leucine, brain 76, 99
lipid metabolism, brain and lead 78, 80
lipid peroxidation 79
lipofuscin 79
litharge 3
locomotor activity 62, 64
lysosomes, brain and lead 78

macrophages 51
malondialdehyde 79
maximal electroshock seizure 68
maze see also individual mazes, acquisition
    performance 120
medical uses 9–11
mental impairment 25–7
  lead in drinking water 29
  retardation 28
3-mercaptopropionic acid, clonus and
    lead 61, 71
metabolic effects 12, 13
metabolism 17–31
α-methylparatyrosine (MPT) and lead
    effects on behaviour 60, 69
methylphenidate and lead effects on
    behaviour 60, 69
microglia and lead effects 106, 115
midbrain 50
milk, lead exposure and behaviour 62, 63
  lead source 37, 62
minimal brain dysfunction 28
  animal model 124
mining
  ancient 8
  Romans in Britain 9, 80
mitochondria 51
  phosphorylation, brain 77
  vacuolation 113
mitotic index, cerebellum and lead 120
monoamine oxidase 87
monoamines and hyperactivity 28
morphine catalepsy and lead 72
morphometric studies 128
motivation 73
motor activity 54
motor coordination 55, 63
Mueller cells 120
mutagenic properties 23
myelin
  accumulation reduction 108, 115, 116

  damage 102, 103
  reduction in encephalopathy 115, 116
myelination 115, 116
α-naphthyl esterase 78

neostigmine
  action 71
  lead effects on behaviour 61, 71
nerve cell changes and lead 106, 107
nerve conduction studies 102
  velocity 102
nerve damage 102, 103
  lead workers and blood levels 102
neurobiological development 62
  EEG measurements on offspring 62
neurobiological effects 1, 129
neurochemical effects 75–100
  biochemical CNS effects 75–82
neurochemical studies 21
  objectives 127
neurological damage, lead exposure 25–7
  asymptomatic 25
  symptomatic 25–7
neurological sequelae 124
neurone, effects of lead 106, 107, 110–12,
    116–19
neuropathology 101, 103–21, 124, 127
  animals 103–13
  nerve cell changes 19, 20, 101
  vascular permeability 19
neurotoxicity 35
neurotransmitters see also individual
    transmitters and systems
  lead effects 82–99
Ninmah 7
noradrenaline, brain levels and lead 88,
    95, 97

occupational risks 13
oedema, brain 19, 35, 101, 104, 105, 109
  components 113
  sodium 82
oligodendrocytes, lead and microglia 106,
    110, 111, 115
open field behaviour 54, 55, 62
operant task performance 63
  differential reinforcement, low rate 58,
    69
  fixed ratio response and lead 58, 67
    blood levels 67
    brain levels 67
    dose-response 67
  ratio and interval 66
  reinforcement 64, 66
  schedule reinforcement 66
  temporally-spaced responding 58, 67, 68
Ophereth 7
optic nerve myelination 108, 115, 116

ores 2
organolead production 4 *see also*
    tetraethyl, tetramethyl lead
orientation and lead dose in rats 57, 65
output routes 17–19
oxidative phosphorylation 81

paraplegia, neonatal rats 36
pentylene tetrazol convulsions and lead 61,
    71
peptidergic systems 83
peripheral nervous system, lead
    morphological effects 101–3
peripheral neuropathy, lead-induced damage
    13, 102, 103
petechial haemorrhage 104, 114
petroleum, lead content, legal restrictions
    14
pharmacological challenge 68
phenobarbitol, lead-treated behaviour 71
phenylalanine, lead effects 76, 99
phenylethanolamine-*N*-methyl transferase
    97
phospholipids and lead 78
photoactivity cage 55
photocell chamber 56
photoelectric cage 56
physical properties 2
physostigmine
    enzyme effects 71
    lead effects on behaviour 60, 71
pica 24, 25, 27
    definition 24
picrotoxin and convulsions with lead 61,
    71
pigments 3
    ancient uses 8
pinking 4
piping 12
placenta
    lead damage 22
    lead passage 22
plumbosolvency 28
plumbum (Pb) 2, 9
poitou colic, epidemics 11
pons-medulla 50
porphyrin synthesis 81
presynaptic dense projections 107
prolactin levels 87
proline, brain 77, 99
protein, brain and lead 76, 80
Purkinje cell
    body size 111, 116
    damage 111, 112, 116, 117
    parameters 111, 112
    pyknotic nuclei 116

radial maze 57

radial 8-arm maze 125, 126
    hippocampus 126
    performance and lead 57, 64, 66
reactive gliosis 35
rearing, lead response 55, 62
receptor activity
    dopamine 94, 96
    and lead 69
    muscarinic 71
red lead 3
    uses 8
residential maze 55, 62
respiration, ADP-dependent in brain and
    lead 77, 80
retina, photoreceptor changes and lead
    108, 120
retrospective studies 27–9
reviews
    government agencies 1
    non-government agencies 1, 2
righting reflex 54
RNA, brain and lead 75, 76, 80
Rome, lead poisoning and fall 9
rotarod 55, 63
Rotherham incident 29
running wheel 55, 62

sapa 9
saturn 2, 7
Saturnism 2
Schwann cell degeneration 102
sciatic nerve myelin 103
screening programmes 24
seizure response and lead 59, 68
selective activity meter 55
serine, brain 77
shock elicited aggression 59, 63
shuttle box 57
sialic acid and lead 78
silver
    byproduct 3
    lead association 5
single level response 58
Skinner box 64
smelter, effects of proximity 29, 30
soil lead 30
solubility, lead and salts 4
spatial alternation 57
specificity, animal studies 123
spermatozoa, lead effects 23
sphingomyelin 79
spiperone specific binding
    lead 88, 89, 96
    receptors 96
spontaneous alternation 57
    amphetamine 64
    lead dose in rats 55, 63, 64, 66
    measurement 125

startle response   54
still births, lead workers   22
striatum   50
strychnine and convulsions with lead   61, 71
subclinical effects   24, 34, 121
   intoxication   46
substantia nigra, lead inclusion bodies   51
sulphatides and lead   79
sulpiride
   lead effects on behaviour   60, 69
   receptor binding   87, 96
   stereospecific binding and lead   87, 96
surma   8
swimming   54
synapse changes, lead dose   107, 118, 119
synapse density   106, 107, 118, 119
synaptic cleft width   107
synaptogenesis   116, 118, 128

teeth, lead levels and environmental
      lead   30
temporally spaced responding   67
teratology
   behavioural   22
   factors   22
   sources of lead   21
tetraethyl lead   4
   first use   14
   prohibition   14
tetrahydrobiopterin and lead
      metabolism   21, 89
tetramethyl lead   4
T-maze   57, 63, 65
tonic seizure   59
transamination   80
transcorneal ECS   59
tricarboxylic acid cycle, brain   77, 80
trilead tetroxide, red lead use   3
tyrosine, brain   76, 99
tyrosine hydroxylase   94, 95

umbilical cord blood, lead   22
uncoupling effects   80

undernutrition   72, 80, 103, 121
   hyperactivity   126
   lead and amphetamine effects   70
   lead dose   113, 114
   lead interaction   45, 46
   non-specific lead effects   123, 124
   toxicity index   70
uptake
   dietary   17, 18
   inhalation   17, 18
   neonates, macromolecules   36
urinary incontinence   103
urorectocaudal syndrome   23
uses, lead   3
   historical   5–14

vaginal opening   63
vanillyl mandelic acid (VMA)
   brain and urine levels with lead   87, 95
   mice and childrens' urine   21
vascular effects, high-dose lead   105, 106,
      113–15
vascular strands   113
visual discrimination, impairment in
      animals and lead   56, 65, 66
visual evoked response, lead exposure   59,
      68

Wallerian degeneration   102, 103
water pipes, lead in Glasgow   28, 29
water purity, lead pipes   9
weight change and lead in animals   36, 45,
      53, 83
white lead
   manufacture   13
   middle ages   11
   paint   23
   poisoning   9
   uses   3, 8
wine storage and colic   11

Y-maze   63